T5-AGA-408

Social Learning in Technological Innovation

Social Learning in Technological Innovation

Experimenting with Information and Communication Technologies

---

Robin Williams

Professor of Socio-Economic Research, The University of Edinburgh, UK

James Stewart

Senior Research Fellow, The University of Edinburgh, UK

Roger Slack

Senior Research Fellow, The University of Edinburgh, UK

Edward Elgar

Cheltenham, UK • Northampton, MA, USA

Published by
Edward Elgar Publishing Limited
Glensanda House
Montpellier Parade
Cheltenham
Glos GL50 1UA
UK

Edward Elgar Publishing, Inc.
136 West Street
Suite 202
Northampton
Massachusetts 01060
USA

A catalogue record for this book
is available from the British Library

Library of Congress Cataloguing in Publication Data

Williams, Robin, 1952 Nov. 13–
Social learning in technological innovation : experimenting with information and communication technologies / Robin Williams, James Stewart, Roger Slack.
p. cm.
"This book presents the main findings of a major European study of Social Learning in Multimedia (SLIM)"—Acknowledgments.
Includes bibligraphical references and index.
1. Technological innovations—Social aspects. 2. Information technology—Social aspects. 3. Social learning. 4. Technological innovations—Social aspects—Europe—Case studies. 5. Information technology—Social aspects—Europe—Case studies. I. Stewart, James, 1968– II. Slack, Roger, 1965– III. Title.

T173.8.W549 2005
303.48'33—dc22

2004058350

ISBN 1 84376 729 5

Printed and bound in Great Britain by MPG Books Ltd, Bodmin, Cornwall

Contents

List of Case Summaries

List of Figures

Foreword

SOCIAL LEARNING AND THE PROCESS OF INNOVATION IN INFORMATION AND COMMUNICATION TECHNOLOGY

This study shows that the eventual uses and utility of Information and Communication Technology (ICT) applications are often far removed from the initial presumptions of its developers. *Social learning* is therefore crucial to how generic ICT capabilities are applied and used in particular settings. In creating new ICT products and services, diverse players are forced to collaborate: suppliers of ICTs and complementary products, media specialists and users. Certain actors play a key role as *intermediaries*, maintaining such collaboration and knowledge flows. The importance of social learning is reflected in the proliferation of *digital experiments:* pilots, feasibility studies and trials, which provide a forum for resolving the uncertainties and differences surrounding the development of new ICT products and services. ICT supplier offerings remain inherently *experimental*. However, the importance of this dispersed innovative effort, and the knowledge it throws up, has often been overlooked in the face of the compelling technology supply rhetorics, portraying existing offerings as already finished solutions to social needs.

The first part of the book addresses the relationship between the design of technology, its implementation and use, and the crucial social learning processes that link these. It draws attention to the various options for organising social learning, from user-centred design, to evolutionary models in which technical and market development go hand in hand, and *laissez-faire* approaches in which users configure standard commodified technical components to their particular purposes. The analysis highlights the difficulties of building an adequate representation of the context and purposes of technology users and embedding them in ICT design – given the diversity of technology users and the fluidity of their requirements. ICT supplier offerings thus remain an 'unfinished' technology, which evolves in its implementation and use (*innofusion*). Non-specialist 'users' play an active role in fitting these offerings to their purposes, making them useful and imparting meaning (*domestication*).

RETHINKING INNOVATION MODELS AND TECHNOLOGY POLICY PERSPECTIVES

The second part of the book explores the implications for technology policy, highlighting the shortcomings of existing models. The key challenge posed concerns the need for a shift in the focus of technology policy from technology development towards the *appropriation* (innofusion and domestication) of ICTs.

Public support for the appropriation of ICT should include provision for the development of *cultural and information content*, and for a creative effort in implementing and using technologies, and for the dissemination of these experiences to other appropriators and to future technology supply.

The social learning perspective draws attention to the transferability of results and how best to utilise the experiences gained in experiments. The lessons learnt may be contingent and difficult to communicate and generalise. It may not be helpful to search for best practice exemplars: attribution of success or failure is often contested and uninformative, and there are many valuable lessons in projects formally defined as 'failures'. Knowledge about change processes provides a more reliable basis for transferability than seeking mechanistic correlations between prior circumstances and outcomes. A key question is whether the players involved in an experiment are motivated to apply the experience gained more broadly. Public support provides crucial resources – but needs to be carefully configured to avoid unhelpful outcomes (e.g. where funding favours launching new projects over exploiting existing products and building markets).

Acknowledgements

This book presents the main findings of a major European study of Social Learning in Multimedia (SLIM), funded under the Targeted Socio-Economic Research programme of the European Commission Fourth Framework Programme (Contract 4141 PL 951003 May 1996–January 1999). This eight country study analysed the development, implementation and consumption of the multimedia technologies that are expected to underpin the transition to an information society.

The SLIM Research Consortium was led by Professor Robin Williams, Director the Research Centre for Social Sciences/Institute for Studies of Science, Technology and Innovation at the University of Edinburgh (UK), with research assistance from Dr Roger Slack and James Stewart.

The other SLIM research centres and key researchers were:

- Cellule Interfacultaire Technology Assessment (CITA), Faculte Universitaire Namur de la Paix (FUNdP) (BE); Clare Lobet-Maris with Beatrice van Bastelaar[1]
- Department of Technology and Social Sciences, Technical University of Denmark (DK); Christian Clausen, Finn Hansen, and Birgit Jaeger[2]
- The Telecommunications Research Group, University of Bremen (D); Professor Herbert Kubicek with Ulrich Schmid and Bernd Beckert
- Communications, Technology and Culture Research Centre (COMTEC), Dublin City University (IRL); Paschal Preston with Aphra Kerr and Stephanie MacBride
- Faculteit der Cultuurwetenschappen, Maastricht University (NL); Professor Wiebe Bijker with Harro van Lente, Mark van Lieshout and Tineke Egyedi
- Senter for Technologi og Samfunn, Norwegian University of Science and Technology (NO); Professor Knut Sørensen with Jarle Brosveet, Hede Nordli and Hendrik S Spilker,
- Institute for Research on the Built Environment (IREC-EPFL), Ecole Polytechnique Federale de Lausanne (CH). Pierre Rossel with Olivier Glassey

A large number of specific studies have been undertaken and many have been published elsewhere. This book seeks to develop an overarching understanding. We have drawn freely upon the unique empirical and conceptual resource provided by the case studies and sectoral analyses developed by our colleagues in the SLIM network. Special mention is needed of the contribution of Knut Sørensen, particularly in developing the concepts of domestication and social learning. We have drawn upon his exegesis in a way that goes far beyond that which can readily be conveyed by the normal methods of citation (particularly in Chapter 3). Though responsibility for writing this book (and any failings) rests with the named authors, we must emphasise the equally important inputs from our collaborators.

NOTES

1 With additional assistance from Jean-Claude Burgelman at the Free University of Brussels, Yves Punie and Frank Neuckens.
2 Now at Roskilde University.

PART A

Social Learning: Understanding the Process of Innovation in the Application of ICT

1. Introduction

THE CONTEXT OF THE STUDY

This book examines a period in which the installation of 'information superhighways', coupled with advances in processing power and usability of information technology, including notably the Internet and the World Wide Web, had led to widespread expectations of the rapid adoption of a new cluster of Information and Communications Technologies (ICTs) (Kubicek, Dutton and Williams 1997, Slack, Stewart and Williams 1999). Moreover, applications based upon these technologies – variously described as digital media and 'multimedia' products and services etc. – were expected to be widely diffused in many areas of working and social life, and to have profound social and economic implications. In short, these technologies were expected to underpin the transition to an 'Information Society'.

The rhetorics of the Information society have become increasingly compelling and pervasive, and have come to inform public policies. However, the strength of these predictions of technological and social transformation hides the lack of certainty about the way ICTs will be applied and used, and thus the character of the Information Society, that can be expected to emerge.

How can we assess the prospects and societal implications of these new technologies? Experience with these emerging technologies is still rather limited. Much contemporary discussion is based upon *visions* of future offerings – which are predominantly *technology-driven* visions and rhetorics, and informed by the perspectives of suppliers and a technically focused community of practitioners and pundits. However, history shows us that what eventually emerges is often far removed from these technically driven visions, in terms not only of the rate of uptake of technologies, but also, more profoundly, of the kinds of technologies that become widespread and their uses and social and economic outcomes (Bruce 1988, Fischer 1992, Dutton 1995). It also shows us that in some instances change can be far more rapid and far reaching than expected, the fax[1] and the World Wide Web being cases in point. Socio-economic research into the development, implementation and use of existing and emerging technologies can make an important contribution here.

This book reports the overall findings of a major European study undertaken of Social Learning in Multimedia (SLIM). Research teams from eight countries analysed the development, implementation and consumption of the ICT applications that are expected to underpin the transition to an information society. The study draws insights from the 'social shaping of technology' perspective and other recent advances in technology studies (MacKenzie and Wajcman 1985, Williams and Edge 1996, Sørensen and Williams 2002) which flags the choices between different options available at every stage of technology development, and the interplay between 'technical' and 'social' factors influencing which option is selected. Another important intellectual resource is from cultural studies with its detailed research into the consumption of artefacts (Silverstone and Hirsch 1992). Building upon these, we deploy the concept of 'social learning' (Rip, Misa and Schot 1995, Sørensen 1996) to highlight the active role of a variety of players – technology users as well as developers – who engage with technologies, and with each other, and in so doing reshape artefacts, and transform our understandings of how these may be used and their significance for society.

The SLIM project was carried out at a critical time – when these key ICT applications were beginning to take off in many areas of life. It represents one of the most large-scale and in-depth studies to date of empirical developments in the *creation and adoption* of new ICTs. The project sought to throw more light on these developments and to engage critically with some of the key intellectual frameworks and policy debates in the field. We were thus able to subject to critical scrutiny some of the widespread presumptions that have been made about this imputed multimedia revolution.

The structure of this book

The first part of this book, Part A, presents and analyses findings in relation to the design and development of new ICT products and services, which constituted the main focus of the SLIM research project. A further concern of the SLIM study, which is reported in Part B, was to develop a deeper understanding of the implications for public policies and the strategies of practitioners.

The first part of this chapter seeks to explain the thinking that underpins our thinking – and the turn to the idea of social learning. We then go on to review, briefly, the origins and development of the 'social shaping of technology' perspective. These ideas may be familiar to many readers from a technology studies background. They may wish to turn directly to Chapter 3 where we introduce and develop the concept of social learning in more detail.

RETHINKING TECHNOLOGY POLICY: COUPLING INNOVATION AND SOCIETY

There has been much recent debate about the social and economic implications of new technologies, and about the public policies and business strategies needed to promote beneficial technological advance. Traditional 'linear models' which focus narrowly on generating novel technology artefacts and knowledge are no longer seen as sufficient to achieve successful innovation and thus the creation of new wealth and jobs, let al.one broader social goals such as a better quality of life. The technological systems of modern society are becoming more complex and elaborate in terms of both their 'technical' functioning and their societal use. Moreover, as technologies become more pervasive and central to society, we come to expect more of them. These factors, coupled with the growing dynamism and pace of social and technical change, add to the difficulties of technology planning and management. Some have suggested that we are finding our way towards a 'third phase' of Research and Innovation Policies in which the key challenge is *to couple innovation and society* – combining social objectives with priorities for the dynamics of innovation (Caracostas and Muldur 1998).[2]

Many of these points apply with special force in relation to ICTs. Earlier technology policies that focused upon promoting the *supply* of new technologies had not delivered all the economic and social benefits that had been promised. There was a gulf between people's everyday experience of ICTs and the confident predictions and compelling visions, articulated by technology suppliers and pundits, that ICT would transform our lives for the better (Forester 1989, Dutton 1996). In this context, 'meeting the needs of the user' became a slogan for the IT industry and public policy.[3] There was, however, little understanding of how to match emerging technical capabilities to potential user need. This was compounded, on the one hand, by the increasing dynamism of technology supply at a global level and, on the other hand, by the deep uncertainties surrounding the application and use of these technologies – which were becoming generic in their application across all industrial sectors and, with the falling costs and increased usability of ICT, were expected to become ever more pervasive in their potential adoption in work and in many areas of everyday life.

Linking new technologies to the 'market' of potential users poses considerable challenges, as an accelerating rate of change in core technologies is associated with enormous turbulence in product markets. The falling costs of computing and communications has opened up the possibility of new kinds of products and services. But what will these look like? There has been considerable uncertainty about which products will prove sufficiently attractive to consumers to be commercially viable – particularly given expectations that radical technological change would both allow very different ways of meeting user needs and lead to the creation of new markets. In the case of radical innovation, where there is no existing customer base, knowledge of earlier products and user markets may no longer be a reliable guide. These uncertainties were given greater salience by the potentially enormous costs of research and development of new knowledge-intensive technological systems. In industries such as ICT, these uncertainties regarding future markets and competitive challenges combined. Technological changes have tended to cut across existing product markets and industrial sectors. The brief history of this sector has been characterised by the displacement of many established players by new market entrants. Correctly anticipating – and shaping – evolving markets and user needs are seen as key to the commercial survival of firms in these sectors (Myervohld 1999).

These uncertainties regarding 'user acceptance' of new technologies have become more salient in recent years because technologies such as ICT are being taken up by an ever growing share of the population. Greater scale increases the potential development costs and the commercial benefits of a successful systems. It also increases the difficulties faced by suppliers in knowing their user market as it becomes larger and more diverse. Furthermore these systems are becoming part of everyday life – at work, in the community and in the home – for more people and in a much more direct way than hitherto. For example, over the first four decades of the application of computers in administration most workers had only indirect experience of computing. As technology users moved out beyond the narrow realm of technical specialists, computers became subject to a range of new and potentially diverging demands, from managers and various groups of 'end-users' at work; from family members using them for entertainment and personal activities in the home. User demands have become more critical. It is no longer sufficient to generalise about 'the user'. Although some user needs are generic (for example regarding the physical and perceptual ergonomics of the computer interface), other needs may be more specific. Indeed, ICT applications are increasingly bound up with *social practice*. 'Industry-specific' applications must support the competitive strategy and coordination processes of the organisation – they are 'organisational technologies' (Smith and Wield 1987), replete with information about the specific organisation and its sectoral etc. context. ICT applications in everyday life are subject to potentially more diverse and challenging user expectations, not just functional requirements of productivity and cost-effectiveness, but also complex personal goals relating to individual and group identity and satisfaction.

Social learning – coupling innovation and society in ICT

These kinds of issues of user acceptance and the establishment of new markets have come to be seen as central to the prospects of ICT, in a context in which the pace and dynamism of technological innovation have reached an unprecedented level. Indeed, leading ICT companies like Microsoft have come to see user responses as a crucial, albeit difficult to predict, factor shaping their success, indeed survival, in the turbulent market for information and communications technologies and services (Myervohld, 1999).

These observations point to the need for a more sophisticated and intricate understanding of the relationship between technology and society. In particular, new concepts are required that recognise the importance of this local innovation effort, and the active role of 'users' with their own knowledges and concerns.

To this end we have developed the concept of *social learning*, building upon earlier analyses of the social shaping of technology (debates which are covered in the next section).

From studies of the social shaping of industrial technologies, we deploy the concept of *innofusion* (Fleck 1988) to show that innovation does not stop when artefacts emerge from technology supply but continues (in the so-called 'diffusion stage') as they are implemented and used. Considerable innovative effort may be needed to get technologies to work and prove useful in particular user circumstances. In addition, from studies of the adoption of technologies in the household, we advance the concept of *domestication* to refer to the creative ways in which final-users incorporate artefacts within their local practices, purposes and culture. Domestication here carries the sense of *taming* something which was hitherto perhaps wild or alien, rather than referring specifically to the household (Silverstone and Hirsch 1992; Lie and Sorensen 1996).

Through innofusion and domestication, novel technologies can be adapted to local contexts; incorporated within particular technical configurations and practices of technology use (and, through these processes, acquire their meaning and social significance). These are two closely related facets of the way in which the more or less generic technical capabilities, arising from technology supply, are *appropriated* (du Gay et al. 1997) – coupled to current and emerging social needs. The *appropriation perspective* being developed in this study emphasises the active role of technology users and other local players, in interaction with supply-side players, in establishing new uses of technology and in creating new markets. This perspective focuses upon usage and the practices surrounding artefacts; it emphasises the information and cultural content of artefacts (especially with ICTs) as well as their more technical characteristics (e.g. hardware and software).

The appropriation perspective extends the range of locales and players seen as significant in shaping innovation and its technical and societal outcomes, with potentially important policy implications (Russell and Williams 2002a, 2002b). The focus of traditional technology policy on technology supply produces a 'problem of diffusion', characterised in terms of transforming technological R&D into marketable innovations; getting these innovations implemented and making them profitable (Sørensen 1996). In contrast, paying attention to the related appropriation processes of implementation (and innofusion), consumption and use of artefacts suggests that measures geared towards the technology supply chain may usefully be supplemented by an effort to support *technology appropriation*. In particular the concept of social learning puts the user and use at centre stage and integrates technology development with societal uptake – flagging how members of a society learn about technology offerings and how designers and developers learn about users and uses. When one acknowledges the need for creativity in order to be able to gainfully employ new technologies, one discovers the need to support and stimulate, but also regulate, this creativity. However, this also demands a better understanding of (and greater concern for) what users do.

We have developed the concept of 'social learning' to explicate the detailed mechanisms and processes involved. The social learning perspective first highlights the importance of practical local activity and knowledge – of *learning by doing* (Arrow 1962) – both in developing multimedia systems and in developing usages of multimedia. It second draws attention to the processes of *learning by interacting* (Sørensen 1996), whereby such locally acquired knowledge may be transferred and applied in other contexts. However, to be intelligible and useful in other contexts such knowledge cannot simply be transported – it must also be translated, combined with other knowledges and transformed (Latour 1986, Callon 1987).

One of the important kinds of learning by interacting involves supplier–user linkages in developing new products. Their contribution to successful innovation has been highlighted by evolutionary economic theorists (for example Andersen and Lundvall 1988) through the concept of 'the learning economy'. Social learning seeks to explore empirically and in detail the operation of the learning economy, to analyse how it is shaped by its particular socio-economic context and the strategies of the actors involved. We are thus interested to explore social learning as a process of negotiation, subject to conflicts of interest amongst players with rather different capabilities, commitments, cultures and contexts.

This is one of the ways in which the social learning perspective seeks to enrich analysis of the social shaping of technology, and to explicate its policy implications. We return to these questions in detail in Chapter 3.

The remainder of this chapter seeks to convey briefly the main precepts of the social shaping of technology perspective and how they have been developed and how they have been applied to analysing ICTs.

THE SOCIAL SHAPING OF TECHNOLOGY

Our study calls into question many received views about ICT and technical change more generally. It is therefore useful to consider the changing analyses that have emerged of the relationship between technology and society, and the emergence of the social shaping perspective as a reaction against traditional understandings of technology. As this analytic tradition may not be familiar to many of the intended audiences of this book, outside the field of science and technology studies, we briefly review this debate. These ideas underpin the analysis we develop around the concept of social learning. This is elaborated in Chapter 3 where the main theoretical developments involved in this study are to be found.

SST and the critique of traditional deterministic accounts

The 'social shaping of technology' (SST) perspective, which informs this study, emerged through a critique of traditional 'deterministic' approaches which took for granted the character and direction of technological advance and thus limited the scope of socio-economic inquiry to addressing its outcomes or 'impacts'. Under the rubric of the social shaping or 'social construction' of technology, scholars from a wide range of backgrounds have broadened the agenda by analysing the *content of technology* and the particular *processes involved in innovation* (MacKenzie and Wajcman 1985, Bijker and Law 1992, Williams and Edge 1996). Their research highlights the range of social as well as technical factors – organisational, political, economic and cultural – which pattern the design and implementation of technology and which thus also shape its social implications.

Mainstream 'technocratic' approaches to technology inherited from post-Enlightenment traditions a concept of technological progress that did not seek to problematise the process of technological innovation. In its deterministic versions, technology was seen as having its own self-evident dynamic – reflecting a single (for example technical) rationality, or an economic imperative. Alternatively, instrumental views saw technology as readily available and capable of being rationally applied to meet social and economic needs – for example, the dominant neo-classical tradition of economic analysis, with its assumptions that technologies will emerge fluidly in response to market demands (Coombs, Saviotti and Walsh 1987). Either way, technological progress could be taken for granted.

These views of technological progress were not well equipped to deal with the experiences of technology, which have been seen as increasingly problematic since the 1970s, on at least two fronts. First is the experience of unintended and undesired consequences of technology (for example health and environmental hazards). Second, the growing pace and salience of technological change has drawn attention to the difficulty of achieving successful technological innovation – in terms of generating new technological knowledge and applying it; fitting it to existing and emerging demand. The latter concern was given impetus by the growing realisation that the traditional approach of supporting technological supply was not, by itself, sufficient to achieve technological advance, let al.one its application to achieve improvements in economic performance and social well-being. These experiences prompted research into the relationship between technology and society – including SST research. In contrast to the certainties held out by rhetorics of social and technological progress, technological change was revealed as a highly uncertain process. A critique of this simplistic and uni-dimensional interpretation led to the espousal of more complex, 'interactive' models of technological change that highlighted the influence of a range of players involved in the use of technology as well as just in technological supply.

SST criticised the 'technological determinism' inherent in such traditional views of technological progress, as exemplified in ideologies of 'technological imperative', prevalent in the early 1980s, which suggested that particular paths of technological change were inevitable. This involved two presumptions: that the nature of technologies and the direction of change were unproblematic or predetermined (perhaps subject to an inner 'technical logic' or 'economic imperative'), and that technology had necessary and determinate 'impacts' upon society – technological change thus produces social and organisational change (Edge 1988).

Instead, SST studies showed that technology does not develop according to an inner technical logic but is a social product, patterned by the conditions of its creation and use. A variety of technical options are available at every stage in both the generation and implementation of new technologies; which of these is selected cannot be reduced to simple 'technical' considerations, but is patterned by a range of broader 'social' factors.

Central to SST is the concept that there are 'choices' (though not necessarily conscious choices) inherent in both the design of individual artefacts and systems. If technology does not emerge from the unfolding of a predetermined logic or a single determinant, then innovation is a 'garden of forking paths'. Different technological routes are available, potentially leading to different outcomes in terms of the *form* of technology: the content of technological artefacts and practices. Significantly, these choices could have differing implications for society and for particular social groups.

The general thrust of SST is to problematise and open up for inquiry the character of technologies, **as well as** their social implications. We can analyse social and economic influences over the particular technological routes taken (and their consequences). SST thus goes beyond traditional approaches, that were merely concerned with assessing the 'social impacts' of technology, to examine what shapes the technology which is having these 'impacts', and the way in which these impacts are achieved (MacKenzie and Wajcman 1985). SST research could, it was hoped, identify opportunities to influence technological change and its social consequences, at an early stage – moments at which accountability and control could be exercised.

SST approaches attempt to grasp the **complexity** of the socio-economic processes involved in technological innovation. In contrast to the presumption that technology supply would generate solutions that corresponded to user requirements, that could then be simply diffused to fulfil society's needs, SST has shown that identifying, let al.one meeting, current and emerging demand for technologies can be difficult. Indeed, a key challenge for technological innovation concerns how to match new technical capacities to 'social need'. This is a potentially difficult problem because social needs, and the means by which they may be fulfilled, are not fixed entities, but evolve, partly in the face of new technical capabilities. SST has contributed to the rather belated recognition of the importance of 'the user' in technological change over the last decade, as well as to the development of an 'interactive' model of innovation. In contrast to traditional 'linear' models, which saw innovation as restricted to technology supply, the interactive model emphasises the innovative effort as technologies are implemented, consumed and used, and highlights the interaction between technology supply and use.

Consideration of (intermediate and final) users draws our attention to the *range of players* involved in innovation, with their different relationships to the technology and varying commitments in terms of past experience and expertise. These players – including technical specialists and other groups from supplier organisations, policymakers, existing and potential users – may have widely differing understandings of technology and its utility. The character and utility of technologies are not defined by any one player but emerge through a complex process of *negotiation* between diverse players. The complexity of these interactions is one of the reasons why **the development and application of technology involves deep uncertainties**. Technological innovations often fail altogether; they usually develop far slower than suppliers and promoters predict, and may follow rather different trajectories than was initially anticipated. Similar uncertainties surround the outcomes of innovation – which mean that it is extremely difficult to predict the 'socio-economic impacts' of technologies with any useful accuracy. One reason for this is that these outcomes are not inherent in the artefacts, but relate to the ways these are applied within particular social settings. So even where the *design* of artefacts embodies particular values or the objectives of particular actors, there is still scope for others involved to achieve some flexibility in their *implementation and use*. The implications of technology for society are therefore rather complex and frequently involve unintended and unpredicted outcomes. Given this situation, it is hardly surprising that attempts to promote or to control technology have not been very effective to date.

In seeking to capture this complexity, SST has developed various sets of concepts which address first the *dynamism* and turbulence of technological innovation, highlighting the *negotiability* of technology (Cronberg 1992); the scope for particular groups and forces to shape technologies to their ends, the *flexibility* of technology (i.e. its openness to renegotiation by local actors) and the possibility of different kinds of ('technological' and 'social') outcome. Second, it seeks to explore the extent and manner in which technological choices may be foreclosed. SST points to processes of *closure* – the ways in which innovations may become stabilised (Pinch and Bijker 1984) and of *entrenchment*, through which particular technological options become generalised. Entrenchment underpins the cumulative development of technological knowledge and increasing returns to established technological options as a result of earlier 'sunk' investment. Entrenchment is critical to the success of a technology. However, it may, equally, result in 'lock-in' to particular options that may become outdated or have undesirable consequences. Such entrenchment is countered by the dynamism, both of new technological capacities and of 'society', which may undermine and destabilise existing expectations and development paths. SST has begun to explore this complex and contradictory interplay surrounding technological innovation between entrenchment and dynamism.

Innofusion and the critique of linear innovation-diffusion models

Fleck (1988) highlighted the shortcomings of traditional 'linear models' which saw innovation as being restricted to the technology supply side – as if finished artefacts would emerge from the research and development laboratory as 'black-boxed' technical solutions, already corresponding to user needs, that could simply be diffused through the market to potential users. Instead, he advanced the term 'innofusion' to describe the 'processes of technological design, trial and exploration, in which user needs and requirements are discovered and incorporated in the course of the struggle to get the technology to work in useful ways, at the point of application' (Fleck 1988: 3).

The concept of innofusion specifically counters prevalent models of *diffusion*, with their implicit view – inherited from the idea of transport of atoms in physics – of an essentially passive process, in which what is diffused is unchanged and all that was in question would be the rate and equilibrium point of diffusion (Rogers 1983). In contrast Rosenberg (1979) showed that, in technological diffusion: '[t]he diffusion process is typically dependent on a stream of improvements in performance characteristics of an innovation, its progressive modification and adaptation to suit the specialised requirements of various sub-markets, and the availability and introduction of other complementary inputs which decisively affect the economic usefulness of an original innovation' (Rosenberg 1979: 75). In contrast, the terminology of innofusion insists that innovation continues beyond the laboratory as technologies are diffused, implemented and used. Fleck's account focuses on the implementation of industrial technologies. Other work has focused upon the translation (Latour 1986, Callon 1987) and transformation of artefacts as they move from the laboratory to commercial production to widespread adoption. As we see in Chapter 4, homologous concepts have been advanced regarding the active and innovative characteristics of the processes of consumption and use of technologies in other settings – and in particular their domestication within the household (Silverstone and Hirsch 1992).

The importance of innovation around the application and use of technologies is one of the reasons why the potential uses and utility of a technology may not be well understood at the outset of a programme of innovation. The history of innovation is full of such examples (Williams 1997, 2000). This observation underlines the serendipity of the innovation process, in terms of our limited ability to anticipate the development of technological capacities and, more importantly, to pre-conceive future applications/uses. Perhaps the biggest difficulties here surround the responses of the user and the evolution of 'social needs'. The future uses and utility of a new application may not at first be self-evident. Partly this is because of the difficulty of anticipating the outcome of the protracted learning processes involving suppliers and users alike as technologies are applied and used. Partly it arises because people may have widely differing attitudes to a particular technology.

A dynamic model technology development paths, meanings and implications

Though retrospective accounts often tend to take for granted contemporary understandings of new technologies and their self-evident functional benefits, historical and contemporary studies of the development, and uptake of new technologies have highlighted the diversity of perceptions by different groups of technologists and engineers (Hård 1994, Vincenti 1994) and 'users' (Pinch and Bijker 1984) regarding technical benefits, particular design choices and the overall significance of the technology. Perceptions are shaped by the particular circumstances of actors (their histories, experiences of technologies, interests and commitments, knowledges, local cultures). The concept of 'interpretive flexibility' has been deployed (from the sociology of scientific knowledge) to capture the multiplicity of meanings that may be associated with a technology and the scope for the various groups involved to articulate their own definitions (Pinch and Bijker, 1984). This flexibility may be countered by processes of 'closure', in which particular design solutions and interpretations gain ascendancy amongst the array of 'Relevant Social Groups' involved (ibid.).

This framework calls for a revision of traditional conceptions which saw the meanings and significance of technologies being built in to technological design and which consequently portrayed consumers and users as essentially passive. In contrast SST shows that, though designers may seek to inscribe particular intentions into material artefacts, this does not foreclose questions about the meanings and uses of a technology. The final consumer may have little opportunity to influence the design and development of artefacts such as domestic goods, other than the 'veto power' to adopt or not (Cockburn 1993). However, even in this setting it is important to acknowledge the scope for these actors to articulate their own representations of technologies and uses which may differ from those articulated by technology suppliers (Akrich 1992, 1992a, Sørensen and Berg 1992, Cockburn 1993). Consumers have shown remarkable inventiveness in side-stepping design presumptions about the use of artefacts, and adapting technologies to their own purposes. In this sense closure is never final and consumers and users are actors, able to exercise choice and inventiveness in the deployment and use of supplier offerings (Sørensen 1994b). This is not to imply total flexibility regarding the use and outcomes of artefacts – some interpretations and some outcomes are more probable than others. However, the significance of an artefact is not fixed in the laboratory, but is only finally resolved in its implementation and use, as it is incorporated within the practices and culture of the user.

The negotiability of technology

SST, in emphasising the scope for local actors to redefine and impose their own understandings, thus suggests that it is unhelpful to talk of the socio-economic 'impacts' of technology, as if these were somehow inherent in the artefact. Outcomes arise through the interaction between artefact and its social setting of use. Rather than imposing particular uses and outcomes, artefacts offer a range of constraints and affordances in their use. And the fluidity and flexibility in use, which may be 'designed into' many artefacts (especially ICTs) suggests that some artefacts may be associated with an extremely wide range of outcomes.[4]

SST in this way emphasises the *negotiability* of technologies – in the sense that artefacts typically emerge through a complex process of action and interaction between a wide range of players involved and affected – including the various groups of (intermediate as well as final) users, suppliers of complementary as well as competing products, promoters, consultants etc. One corollary of this is that the development and application of new technologies typically requires the bringing together of different kinds of knowledge and other resources – which are often unevenly distributed across different organisations and institutions.

Studies of the development of new technologies and technological systems have emphasised how they emerge through the creation of formal or informal alliances of players with the resources and technical expertise needed, united by certain broadly shared concepts or visions of as yet unrealised technologies. SST has begun to uncover some of the strategies for creating/mobilising such 'sociotechnical systems' (Hughes 1983), 'constituencies' (Molina 1989) 'ensembles' (Bijker 1993, 1994) or 'actor-networks' (Law and Callon 1992), for example, 'enrolling' local players in a broader network (Law and Callon 1992) and 'aligning' their expectations around realisable objectives (Molina 1994).

Though early SST research was primarily concerned with the initial development of new technologies and artefacts, and the socio-technical systems/constituencies of players involved in the generation of new technologies, the interactive model highlights the need to look at whole 'circuit of technology' (Cockburn and Furst-Dilic 1994: 3), from design and production through to consumption and use, to understand how technologies and the social implications are shaped.

The entrenchment of technology

Stabilisation and entrenchment are crucial to the success of a technology. First, the commercial viability of a new technology may rely on achieving a critical mass – or, more accurately, upon convincing a critical mass of players to invest in the technology (Graham, Spinardi and Williams 1996, Schneider et al. 1991, Schneider 2000, Williams 1995). Where technologies exhibit economies of scale and network externalities, uncertainties arise about whether enough people would be convinced to adopt a technology to recoup development costs; to bring down the price of a product and to make it cheap enough to be widely attractive to consumers – both intermediate users (e.g. by establishing a market of content providers) and/or final-users of products and service). If sufficient numbers of other users cannot be convinced to sign up to make a technology attractive to a potential user the technology may not fulfil expectations (e.g. by failing altogether – as in the case of videophones [Dutton 1995] or only developing slowly and only achieving viability in niche markets – as in the case of videotex in the UK [Bruce 1988]). This is particularly important where technologies demonstrate what economists describe as 'positive returns to sunk investment' (the gradual improvements in performance of technologies as technological knowledge is refined and developed) and 'positive network externalities'. The latter refers to technologies such as the telephone or Electronic Data Interchange (Graham, Spinardi and Williams 1996) the benefit of which to existing users is increased by the arrival of additional users. These are often 'network technologies',[5] whose functioning and utility depend not on the individual artefact but on how the artefact is integrated into a wider network of inter-operating elements (for example in relation to new 'delivery systems' and platforms), and which are subject to marked path dependencies and scale effects. These features mean that attempts to create new, or change existing, complex technological systems may be beset by problems and uncertainties surrounding entrenchment. In these settings, markets may fail, at least in the short term, to guarantee the uptake of more productive technologies.

These considerations point to the benefits of 'closure' – which may motivate the players involved to seek agreement about standards and the technological paths to be followed. On the other hand, such rigidity can be counter-productive. The results of earlier technological choices can constrain later technological decision-making (Rosenberg 1994). *Path dependencies* arise from the increasing returns on 'sunk investment' into particular technical standards, given the cumulative nature of technological advances. Such 'path dependencies' can result in 'lock-in' to established solutions and standards, even where these technologies are no longer seen as optimal (David 1975, Arthur 1989, Cowan 1992). Well-known examples of this are the QWERTY keyboard and railway gauges.

Entrenching factors are not just economic, but include, importantly, shared perceptions and expectations of a technology. The success of new ICT products does not simply reflect their functionality and price, but also the extent to which they are compatible with the skills, understandings and habitual practices of potential users (Miles 1990, Thomas and Miles 1990).

Alignment of perception is an important step in innovation. For example, engineers must project visions and align expectations of a technology and its capabilities in order to enrol the support of other players if they are to obtain the technical and human resources needed to create it (Molina 1994). This may involve establishing consensus around particular technological concepts and options. The extension of such consensus to include the range of suppliers, consumers of an artefact and other relevant social groups presages technological 'closure' and the stabilisation of technological artefacts (Pinch and Bijker 1984). However, alignments that are premature or embody particular presumptions or visions of a technology can focus attention too narrowly on particular technological paths in a way that can prove disastrous if circumstances and perceived user requirements change (for example, the abandoned $150 million investment in the EFTPoS UK pilot for Electronic Funds Transfer at Point of Sale [Howells and Hine 1993]). This is just one example of how inflexible development contexts may result in inflexible technological designs (Collingridge 1992). Entrenchment therefore represents something of a two-edged sword.

However, closure is never final (Rosen 1993). The possibility remains of reversing earlier choices (Latour 1988). However, the perceived costs of modifying an earlier decision (in their broadest terms including time, money, uncertainties and the effort of abandoning existing routines and knowledge investments and learning new approaches) may appear prohibitive. *Irreversibility*, is always a complex sociotechnical accomplishment, rather than a technical fact (Callon 1991, 1993, Collingridge 1992). Established facts and artefacts remain perennially vulnerable, in particular, to dynamism in the technological system and broader society.

So we find that technological innovation is subject to two contradictory tendencies. On the one hand, we identify processes forces that will tend to stabilise technologies, by aligning expectations, and reducing the uncertainties and costs of established models. On the other hand, the dynamics of the development both of new technological opportunities and of user requirements – new problems thrown up by societal changes, and the articulation of new ways of linking those problems with technical possibilities –may open up new application possibilities, and undermine existing solutions, reversing the trend to stabilisation (Brady, Tierney and Williams 1992). An important influence here concerns the dynamics of particular product markets. Economic pressures may favour established approaches, in a context of positive returns on past investment and economies of scale. Standardisation of technology has played an important role – yielding substantial economic benefits for various players: bigger markets and greater profits for suppliers and important price advantages for consumers who can share development costs (which may be very high with high-technology, knowledge-intensive products). In this way, some technical artefacts may become stabilised and standardised. They may be made available to the user through the market as 'black-boxed' solutions, as 'commodities' with well-established attributes. However, competitive considerations may also counteract this. Once new markets have been established this will attract new entrants to compete with established suppliers, who in turn may seek to differentiate their offerings and retain their existing customer base through technological leadership. And users may seek to gain 'competitive advantage' by adopting these new offerings. So here again we find a complex interplay between standardisation and commodification – consolidating and undermining technological entrenchment over time.

THE SOCIAL SHAPING OF INFORMATION AND COMMUNICATION TECHNOLOGIES

We have already pointed to the deep uncertainties surrounding the development of ICTs, as new technological capabilities emerge with the accelerating pace of advance in core technologies and become bound up with compelling, technically centred visions of change, only to come into collision with the slow and uneven pace of uptake and the often ambivalent responses of potential consumers.

We have little experience of these new technologies of the information society. However, the past can provide us with important insights into the future (Bruce 1988, Dutton 1995). The lessons from earlier technologies suggest that these early visions may be deeply misleading (Williams 2000). A growing body of research has addressed the social and economic factors shaping the development and use of ICTs and its outcomes (Williams 1995, Dutton 1996, Clausen and Williams 1997, Kubicek, Dutton and Williams 1997, Slack, Stewart and Williams 1999). A recurrent element in this shaping and reshaping of ICTs arises because the uses of technology are not simply an extrapolation from technical capabilities, but emerge through a complex interaction between ICT supply and use. Though it may be resolved differently in different periods and contexts, we can see this interaction recurring throughout the history of ICTs – though the contribution of users to ICT innovation has not been adequately recognised. Thus ICT users have contributed to transforming our understanding of the significance of artefacts, and perhaps in stimulating and redirecting ICT innovation, contributing to the emergence of new technologies. A brief account of our analysis of these features – focusing on the interplay between technology supply and use – will explicate the model we have developed for understanding the character of contemporary ICT artefacts, the process of their innovation and the future prospects of ICT applications.

Let us start by considering why the promises of social transformation and 'revolutionary change' that would arise from the supply of new ICTs have not materialised. The rhetorics of ICT supply suggested that new technological offerings emerging from technology supply would provide finished solutions that could be readily and widely applied to the full range of social activities and problems. ICT supply has certainly been extraordinarily dynamic in terms of the increasing power and falling costs of new products. However, the domain of societal application and use of ICTs has proved more complex and obdurate than many technologists had presumed. There are a number of reasons for this (Williams 1997). For example, some of the difficulties applying ICTs in organisations can be related to the discrepancy between computing – with its roots in formal and mathematical models and representations – and the character of much human communication and decision-making processes based upon complex judgements and interpretation in contexts that are typically characterised by ambiguity and uncertainty and inherently difficult to describe in formal mathematical terms. Successes in early administrative applications of 'electronic data-processing' to routine and simplified information processing activities (e.g. accounting) which could readily be described in mathematical terms, converted to algorithms and implemented in software encouraged technical specialists to underestimate the complexity of application areas, and the consequent difficulties of applying ICTs more broadly.

The claims to universality of ICT, based upon formal mathematical techniques, came into conflict with the diversity and specificity of the social contexts of application and use of ICT occupied by diverse kinds of 'users' with their own practices, purposes and culture. Users are various and may have quite different perceptions of ICT artefacts and their utility. Matching supplier offerings to user need can thus prove problematic, particularly with novel technologies where there are few established models of the application and use. While technology-driven views typically take the utility of the artefact for granted – assuming that new functionalities offered will somehow match user requirements, identifying 'user requirements' remains far from straightforward: users do not have a finite and determinate set of existing 'requirements'; requirements are constructed – built upon earlier templates and evolve as artefacts and practices change.[6]

The logics inherent in new technologies did not drive social and organisational change. Indeed, research into the industrial implementation of ICTs suggested a mutual shaping between technology and organisation, in which, in the short term at least, organisational structures and practices seemed to be more powerful reshapers of technology than the other way around (Webster 1990, Williams 1995, Clausen and Williams 1997, McLoughlin 1999). Rather than the transformation of 'society' by 'technology', we thus find a complex *interplay between the dynamics of technology supply and use.*

Following on from this, the application domain has been the site of considerable experimentation and innovation as users and others struggled to get these generic offerings to work and prove useful in actual contexts of use (Fleck 1988, 1988a). Indeed, *'users' have made an important contribution to innovation in ICT* – even in relation to core technologies which today are seen as the virtually exclusive preserve of specialised IT suppliers (Brady, Tierney and Williams 1992). Thus industrial **users** of automatic data-processing machines, such as Prudential Assurance and Lyons, were heavily involved in the construction of the earliest commercial computers (Campbell-Kelly 1989) and in the commercial application of robotics (Fleck 1988). 'User-led' innovations throw up ideas which may be taken up in future technology supply and ultimately become sedimented in core technologies. Industrial users were involved in the creation of the first operating systems (Friedman 1989). This continues today – the most famous recent example being the development by CERN scientists of the ideas that underpinned the World Wide Web. It is in the area of mass-market ICT products for use in the home and in everyday life that we have some of the clearest evidence about the centrality of user responses in shaping not only the differential success of new products, but the very conception of these products (for example the camera [Jenkins 1975] or the video cassette recorder [Roosenbloom and Cusumano 1987]). Perhaps the most striking and well-researched example is the telephone, which was originally conceived and promoted as a business communication tool for conveying price information to farmers, but which was reinvented by people in rural areas, particularly women, as a medium for social communication (Fischer 1992). The evolution of the home computer equally exemplifies the role of domestic users in appropriating and shaping technologies (Silverstone and Hirsch 1992). Though initially promoted as a means of carrying out various 'useful' activities (word-processing, household accounts and, in particular, as a support for educational programmes), this was largely subverted by consumers in the family. The technology became subject to a web of competing conceptions articulated by various players: government, suppliers, parents, children. Boys in particular, though perhaps pressurising their parents to acquire a computer for educational reasons, were really interested in using them for playing computer games. This has shaped the evolution of home computers, leading to the creation of a specialised market for these products, transforming it into a tool for entertainment rather than productivity (Haddon 1992, Murdock, Hartmann and Gray 1992). Thus user innovation has led to *the transformation of our understandings of a technology*, radically rethinking its uses and significance.

On the other hand, the universality of these mathematised computing technologies, particularly in relation to core components such as memory processing and transmission, has underpinned enormous improvements in technical capabilities, scale economies in production and consequent reductions in the cost of processing power. These core capabilities, in common with many knowledge-intensive products, exhibit peculiar economics, particularly where products can be mass-produced at low cost, given the high up-front costs of labour-intensive knowledge acquisition/ origination and the low subsequent costs of reproduction. The economics of certain 'globalised' ICT products such as microprocessors and computer operating systems involves huge research and development costs of launching new products, coupled with massive potential economies of scale if they are successful in the market. This combination brings great uncertainties: huge losses for those that fail and potentially enormous returns for those that prevail. As a result, the production of many core technologies such as microprocessors is increasingly globalised.

Commercial and technological strategies have emerged which try to reconcile the counterposed vectors of the dynamism of supply of new technologies with the stability and obduracy of the market comprising large number of actual and potential users with their more or less specific histories, contexts and commitments. For example, we see the development of 'architectural technologies' (Morris and Ferguson 1993) offering longitudinal compatibility between different generations of technological artefacts, to allow a supplier to innovate some elements whilst keeping the product compatible with the existing market defined by the installed base of complementary products: applications and software (Molina 1992, 1994, Williams 1997).

These interactions between 'global' generic technologies and 'local' user contexts are often played out in relation to software. Software represents the critical layer in ICT systems – it forms the interface between the 'universal' calculating engine of the computer, and the wide range of social activities to which ICTs are applied. For ICT systems to be useful, they must, to some extent, model and replicate parts of social and organisational activity. Software is designed to achieve particular purposes; its design embodies particular values and social relationships (Dunlop and Kling 1991, Quintas 1993, McLaughlin et al. 1999).[7] Particularly in relation to the complex integrated software systems being developed to support the activities of large organisations, software may need to be designed to meet the specificities of organisational practice and structure, and may be sought as a customised solution.

Custom solutions are, however, very expensive. An alternative supply strategy has involved the cheap commodified supply of standardised packaged software solutions. Figure 1.1 (from Procter and Williams 1996) maps out this dichotomy in strategies for the supply of industrial software applications in terms of their volume/variety characteristics between, at the one extreme, custom software designed around the needs of a particular user (highly complex and with a market size of 1) and, at the other, discrete applications (e.g. word-processing software or payroll systems) that can be supplied as cheap commodified solutions to an increasingly globalised market. The former is expensive, but offers solutions that are well matched to local needs; the latter offers very cheap solutions to particular tasks. It has been particularly successful in relation the design and supply of software *tools* – such as spreadsheets which can be adapted by the user to a wide range of purposes – as well as some of the most successful recent ICT applications in communications technologies, such as electronic mail or latterly desk-top video-conferencing, which are in the form of *media*, that offer little constraint on the *content* to be exchanged and that can thus be applied in a wide variety of contexts.

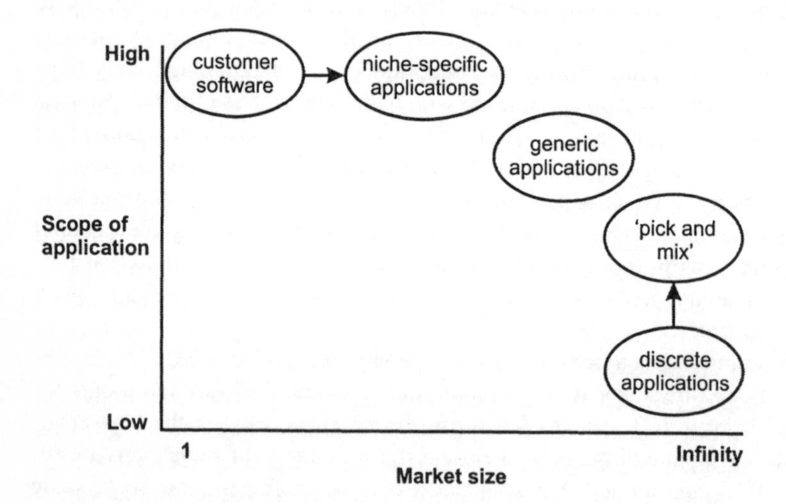

Source: Procter and Williams 1996

Figure 1.1 Volume Variety Characteristics of Packaged Software Solutions

This points to a dichotomy in software development strategies between, on the one hand, attempts to designing more of 'society' (and specific contexts of use) **into** the software and, on the other, attempts to design 'society' (or, to be precise, references that limit applicability to specific contexts of use) **out** of the artefact, and in consumer strategies between maximising utility and minimising price.

The focus of current supply strategies lies between these extremes: for example in attempts to adapt and 'sell-on' complex customised solutions to organisations with broadly similar settings and activities (e.g. credit-card processing) as niche-specific applications; or generic applications designed to be adapted by the user to meet a variety of particular circumstances (e.g. production control and scheduling systems) (Fincham et al. 1995, Webster and Williams 1993).[8]

One other strategy is becoming increasingly influential in contemporary ICT applications: that of configurational technology solutions. With the emergence of inter-operability standards it becomes more easy to select a range of standard components (hardware and software – e.g. personal computers, network management software, database technologies) and to knit them together in conjunction with customised components into a particular configuration that matches the particular circumstances of use. Such *configurational technologies* offer a much cheaper way to meet the particular needs of a complex organisation than fully customised solutions. Complex ICT systems today are increasingly taking on these 'configurational' characteristics. Indeed, given the enormous cost advantages of standard hardware and software components, few developers would build a new system entirely from scratch. The recognition that complex ICT applications emerge not as integrated 'systems technologies' but as 'configurations' (Fleck 1988a) of more or less standardised component technologies and tools is crucial to understanding perhaps the most important factor shaping the evolution of ICT and its uses – the interplay between the dynamism of ICT supply (the increasing power and falling prices, particularly of standardised mass-produced technologies) and the specificity of the social settings in which they are introduced.

The concept of ICT as configurational technology underlines important changes in the character of system development processes, and the growing role of consultants and systems integrators in configuring ICT systems for particular organisations. This model also flags the contribution to innovation of technology users and other players as well as technology suppliers – for example in innofusion in the development of industrial ICT applications. In other settings – for example mass market products for consumption by individuals and their families – there may be less opportunity for final consumers to engage with technology design and development. However, even in this setting it is important to acknowledge the scope for these actors to articulate their own representations of technologies and uses, which may differ from those articulated by technology suppliers (Sørensen and Berg

1992, Akrich 1992 and 1992a, Cockburn and Furst-Dilic 1994). Consumers and local consumer-experts are not powerless in the face of technology suppliers and can take advantage of the configurational nature of ICTs to creating their own technological environment. *Consumption is an active process*, involving selective decisions to purchase the technology and appropriate it within social practices and systems of meaning (Sørensen 1994b).

THE PROCESS OF INNOVATION IN ICT

ICT as configurational technology: the four layer model of ICT innovation

These features help us understand the dynamics of ICT development and application. This is subject to enormous uncertainties (Williams 2000). Whilst continued progress can be expected in the performance of the generic technical components (e.g. increasing information-processing power) there is much less agreement about how these will be configured into delivery systems and about which applications will succeed. As in the case of many other ICTs, there is a tension between the promises held out by rapidly developing technologies and what seems to be delivered (especially in the short term) and between the imputed universal applicability of new generic technical capabilities and the specificities that surround their adoption within various social settings.

The concept of configurational technology is highly relevant to understanding these developments. In particular it provides a schema for understanding how contradictory requirements are reconciled – in particular the tensions between global technological development and its local appropriation, and how supplier offerings may be matched to local user requirements. ICT, like other rapidly changing complex technologies, is markedly *heterogeneous*, combining offerings from a range of different suppliers to meet particular requirements in a way which affords considerable flexibility in development, implementation and use. The concept of technological configurations can be applied at two levels:

1. to highlight the way that a range of *component technologies* (fibroptic cable, microprocessors, software tools) are assembled into particular configurations in terms of *delivery systems* (for example public kiosks, interactive TV or networked personal computers taking these systems into the home);

2. to show how these delivery systems are themselves configured to the purposes of the particular networked ICT *applications* that run upon them (for example teleshopping, on-line newspapers, games, interactive TV services).

Collinson et al. (1996) derived from these considerations a 'three layer model' of innovation in ICT, showing the broad segmentation between components, delivery systems and applications. To this we added a fourth layer to draw attention to the importance of cultural content such as web pages and on-line museum exhibits (Preston 2000). These complementary products could be crucial in determining the viability of a new application or platform – their markets are mutually reinforcing. Successful technology developers have often paid attention to collaboration with content developers (Cowan 1992, Collinson 1993, Nicoll 2000); indeed, content is often the major part of the value chain for a new technology.

The resulting four-level model is summarised in Figure 1.2. The importance of this schema is in charting the complexity of the innovation process underpinning the emerging information infrastructure. In particular, it opens up for examination the level of autonomy between the different levels. It suggests that particular combinations of components can be put together to build a variety of different platforms and applications, with different social and technical characteristics and implications. Conversely, particular components can be transferred from one socio-technical setting into others – as we see, for example, with digital TV set -top boxes being opened up as devices for accessing the Internet. Content may be adapted for specific platforms and delivery systems (given the constraints, for example, of bandwidths, display technology and navigation tools). Conversely, digital content may be ported to other delivery systems.

In terms of the influence of the user and social learning, the key social shaping processes concern the development and implementation of ICT applications and their attendant delivery systems. It is here that the most salient societal choices about the character of the information superhighways and the applications running upon them will be made. For example, some kinds of delivery system will be closely dedicated to particular kinds of application (for example, an interactive television terminal), while others will provide a more open platform on which a range of networked services can be mounted (for example a personal computer linked to the Internet). These choices which affect the flexibility of the delivery system could therefore have important social implications.[9]

	Definition	Examples
CONTENT	*Information and cultural content* *Established forms for this content in terms of media genres and models of interactivity*	Web pages E-commerce catalogues On-line museum exhibits SMS messages Video games
APPLICATIONS	*Specific configurations – services, applications and products, within particular sectors and contexts*	Video clips on demand On-line shopping CD education packages Video-conferencing Electronic cash-cards
DELIVERY SYSTEMS	*Combinations of technologies for storage, display, delivery, distribution*	DVD production+players PDAs Interactive TV /Digital TV systems The Internet
COMPONENTS	*Basic building blocks that can be combined to enable product and system development*	Microprocessors, Video standards ADSL bandwidth Screen resolution Software tools Compression techniques

Source: adapted from Collinson et al. (1996) threelayer model

Figure 1.2 The Four-Layer Model of Innovation in ICT

The significance of these processes is heightened by 'path dependencies', which may be particularly pronounced with network technologies like ICTs. There has been a widespread expectation that particular ICT applications would be the 'key drivers', bringing networked ICT products into the home and workplace, and establishing models for other products and services, which may in turn constrain the kind of further applications that can be mounted upon them. The first widespread application will involve enormous sunk investments, by suppliers and users. This could be an important influence upon the shape of ICT and its societal implications in the medium term, particularly if it leads to the widespread implementation of delivery systems and devices that are dedicated to the requirements of those specific applications (e.g. video-on-demand) rather than offering more flexible configurations (e.g. networked personal computers [Pcs]).

An interesting illustration is afforded by the case of Minitel in France. Though public support for acquisition of videotext terminals was seen as exemplary in creating a market for on-line services, the system has subsequently been bypassed by the emergence and global uptake of Internet-based standards. Difficulties in upgrading the dedicated Minitel terminals to Internet standards constitutes a barrier to access to the World Wide Web by the mass market. Paradoxically, an unplanned consequence of the selection of an industry-standard PC as the terminal for the German videotext system has allowed German users ready access to both delivery systems (Schneider 2000). Our understanding of the ICT platform as a configurational technology also highlights the scope and constraints upon its reconfiguration. One implication concerns the desirability of aligning wherever possible with emerging *de facto* global standards. Other public policy and technology strategy implications may follow. For example, reconciling the desire to benefit from using technologies before closure (and the stabilisation of artefacts and interfaces) with the need to avoid entrenching incompatible solutions suggests the need for strategies geared towards longer-term flexibility, for example, involving *gateways* between different systems and *migration paths* to allow subsequent upgrading of systems to be compatible with emerging standards (Spacek 1997).

Social learning in ICT

Significant technical obstacles remain, for example, regarding the reliability and security of transactions across the Internet. The 'local loop' with its limited carrying capacity continues to be seen as a bottleneck, and much current concern focuses on bringing broadband services into the home. However, we can be rather confident about continued improvements in performance of many of the core technologies. The geometrical growth in computing power suggested by Moore's law is being outstripped by the rate of installation of new telecommunications capacity. A new generation of products and services are being envisaged around the development of digital media, 'multimedia' technologies and the 'information superhighways' that could bring digitised video and sound, as well as text messages, into the home (Collinson 1993, Cawson, Haddon and Miles 1995) and to the emerging mobile and wireless which can bring these to the individual in all the settings of everyday life (Brown, Green and Harper 2002, International Telecommunications Union 2002). A plethora of products are being launched for the workplace (Procter and Williams 1996). Central and local government are investing in developments – particularly in relation to education. Huge markets are anticipated for new products that are interactive, easier to use and more engaging.

However, as this review has shown, there is little understanding at the outset of development of what the products that will eventually prevail will look like, let al.one their social implications (Dutton 1996, Kahin and Keller 1995). Perhaps the most profound uncertainties surround the **societal uptake and use** of the new products and services. This is an area where we have very little direct experience or relevant knowledge. Indeed, one of the key features of 'the digital revolution' in its early stages is that it has primarily been driven by the perspectives of suppliers of equipment and services, informed largely by their expectations of what might be **technically feasible**. This may not be altogether surprising in so far as there were few established exemplars for these future services – so there was little possibility to assess their attractiveness to the user. However, despite the huge amount of energy devoted by suppliers, promoters and media popularisers of multimedia-based products and services, there is remarkably little evidence and understanding about the nature, strength and evolution of user demand. Despite longstanding expectations of a 'killer application' that would herald the digital revolution (Dutton 1996), many of the first-round services launched (e.g. interactive TV) are only on the margins of commercial viability. Some of the early commercial ventures and industrial alliances – driven by particular concepts of the new market opportunities that were anticipated from technology convergence and which in turn fuelled wider expectations about the digital revolution – have already been quietly shelved, often following substantial losses (Fransman 1998).

Many of the successful supplier offerings that have emerged to date are markedly conservative innovations – often simply on-line replicas of existing products and services (Collinson et al. 1996). This may only reflect a very rational strategy by suppliers seeking to minimise uncertainty, drawing on their existing expertise and links with existing customers. However, as we have seen, the experience of earlier generations of ICT suggests that the main products and services that eventually prevail in the future information society may well be far removed from the embryonic offerings available today. It is extremely difficult for suppliers to assess what user requirements will emerge.

Suppliers of new products and services are forced to operate in a context of only limited information and considerable uncertainty about 'the user' and 'their needs' and how they might be satisfied by ICTs. Moreover, 'users' and 'user needs' are not pre-existing and static entities, but emerge and evolve through an interaction between supplier offerings and user responses. The key questions about the future of ICT products and services thus often concern the engagement between supplier and user. Successful applications will doubtless emerge. In the mean time public and commercial promoters have recognised the need to develop better understanding of existing and nascent requirements of potential consumers/users of their offerings. They have even looked to social science for answers as well as launching a growing array of feasibility studies, pilot projects and commercial trials of ICT products and services and related technologies to provide this information. These developments reflect a growing recognition of the importance of 'user' responses to supplier offerings – both the intermediate consumers (for example suppliers of complementary products such as games to run on a platform) and the final consumers of those products and services. Through their veto-power, in deciding whether or not to buy/adopt a new product, and also, as we see later, their selective engagement with supplier offerings, users are not passive recipients of supplier offerings but play an active role in innovation, often using artefacts in ways neither anticipated not intended by the developers, and thus transforming our understanding of the artefact and its uses and significance.

Social shaping of technology research can offer important insights here, as the preceding review has shown. However, it also points to the need for further research and critical analysis. For example, we still cannot provide unequivocal answers to perhaps the most basic question regarding *how best to match user requirements to the new technical possibilities*. For example, debates are continuing about whether the most effective approach to meeting user need will involve building more knowledge about the context and purposes of the user into the design of new applications – or alternatively by designing and launching onto the market generic offerings, and letting final-users learn how to adapt these offerings to their purposes. Though there have been moves to improve the effectiveness and efficiency by which user requirements can be understood and represented in design, it remains the case that most of the key breakthroughs in ICT applications have been to a large degree serendipitous and their success has not been anticipated (as we discuss in relation to such technologies as fax, and most recently Short Message Service on mobile phones [International Telecommunications Union, no date] etc.). Perhaps the more important question concerns what opportunities exist between these two extremes of 'user-centred design' and 'laissez-faire' supply and user-led appropriation? How might these best be combined?

The outcome will depend upon the highly dispersed processes that we describe as social learning – involving a range of social actors: suppliers, policymakers, promoters and intermediate and final-users. The thesis of this study is that society and its various publics can draw more general lessons from studying the creative attempts of these players to resolve particular issues around ICT development, implementation and use.

The experimental character of contemporary ICT applications

When the SLIM research design was developed we anticipated an important focus upon specific 'social experiments' in ICT – perhaps of the sort conducted in Scandinavia in the 1980s (Cronberg et al. 1991). Such social experiments sought to offer exemplary solutions for particular social groups, articulating values and concerns that might be overlooked by conventional commercial developments (for example the needs of economically marginal and socially excluded groups) and based on the involvement of users. One finding from this project is that this particular type of social experiment is today rather rare.

However, the contemporary development of ICT-based products and services remains profoundly *experimental* (Jaeger, Slack and Williams, 2000). On the one hand, government and private sector players feel compelled to become involved and gain experience with the technology; on the other hand, they face deep uncertainties – in particular about the responses of 'final-users' to their offerings, as well as the behaviour of other players – such as competitors and suppliers of complementary products. In this context one of the key ways in which a technology may be carried forward has been through demonstration projects, feasibility studies, pilots and commercial trials (as well as formal social experiments). A plethora of such experiments, feasibility studies and pilots have been launched in most developed (and many developing) countries world-wide, often as part of 'Information Society' initiatives, together with a host of commercial initiatives and trials. Chapter 8 includes a survey of such initiatives across a range of European studies (see also Jaeger, Slack and Williams 2000).

There are important commonalities between these different types of initiatives – particularly in relation to social learning. We will refer to these collectively by the term '*digital experiment*'.

Furthermore, ICT application projects were experimental even where they was not intended to be so (Jaeger, Slack and Williams 2000). Where projects were set up with rather limited goals, their outcomes tended to be rather more open and to offer longer-term opportunities for social learning (even with projects that got into difficulties and failed to meet their objectives) in ways that were not anticipated at the outset. The experimental nature of these developments was not, however, necessarily recognised or properly catered for. The lack of prior attention to social learning possibilities was particularly notable in projects that were launched with the primary objective of developing technical capacities, where there was often very little prior discussion of technology application and use. Of course, even the most 'technical' of trials needed to involve some kind of social application and concept of use. Indeed, such trials were often conducted with 'real life' applications. However, issues around application and use were not always given proper attention. Although experimentation inevitably took place in the course of these trials, these were not necessarily planned for and specified in advance. As we see in Chapter 8, this was particularly noticeable around technology infrastructure developments, such as the Metropolitan Area Networks, which created broadband transmission capabilities around which a swathe of ICT application initiatives arose. In this context, experimentation and social learning around the application and use of ICT are thus smuggled in – as a more or less implicit and covert outcome rather than a central goal of a technically oriented trial. This lack of attention and resources has potentially serious consequences. For example, where projects were only given short-term support, resources were not necessarily available for the longer-term exploitation of experiences (as we see in the concluding chapter). This is not to accept that digital experiments can ever be narrowly 'technical'; these infrastructure initiatives need some kinds of application for testing to take place. Even though formal trials may be couched in terms of a technical feasibility study, this is likely to be the starting point for a more extended informal process of social learning around the development and use of applications that run upon the infrastructure (Jaeger, Slack and Williams, 2000). Though one cannot prevent social learning, in many such *de facto trials* and *experiments* it seems that the lessons are not always being systematically sought or communicated. As a result, it seems, some lessons are lost and have to be learnt time and time again.

Despite the powerful rhetorics of the impending Information society and the digital revolution (and the substantial proportions of gross national product [GNP] being invested in ICTs generally) it is important to remember that ICT application projects – i.e. those involving more innovative technologies or uses of technologies – were often rather *marginal* to the core concerns and activities of many of the key commercial and government players involved in creating new ICT products and services.[10] As a result, the levels of investments were often modest, and it often proved difficult for the proponents to obtain the levels of resources they felt necessary. We can see the resort to this experimental approach as representing a trade-off between competing concerns – to avoid the risk of being 'left out' of a technology that was deemed to represent the future, while reducing the risks of wasting significant sums on inappropriate investments. Whilst the lack of resources and commitments did present problems for the players, it also provided a space for groups within these organisations to experiment with these new technical capabilities.

We see the proliferation of such digital experiments and trials as an indicator both of the importance of social learning in such a novel technological arena and a pointer to the growing recognition of such social learning by many of the players involved. The diversity of these kinds of digital experiment is of particular interest. Beyond these we note the variety of ways in which ICT is being developed and implemented more generally. Digital experiments can be expected to provide a more elaborate and formalised context for social learning. There is, however, little understanding about the actual processes of social learning taking place in such experiments and more broadly. The SLIM project emphasised that we do not know how best to organise such learning processes. The researchers were not committed at the outset to particular approaches such as social experiments specifically, or even formalised digital experiments in general but were instead open, for example, to the idea that conventional models of technology supply through the market might suffice. The intellectual challenge that this project sought to address was thus to conduct an open-minded exploration about how ICT innovation proceeded in different contexts, and the social learning opportunities they afforded.

NOTES

1 Though facsimile transmission was conceived as early as 1843, with its first commercial launch in 1865, this and successive attempts to develop the technology did not meet with success apart from particular niche applications – for example its use by newspapers to transmit photographic etc. images to remote offices – until the 1980s. Its (largely unexpected/unpredicted) explosive growth then was attributed to the success of Japanese suppliers in 'manufacturing a superior machine' that was cheap and designed to be easy to use (Oakley 1990, Coopersmith 1993). Only then was fax technology close enough to the requirements of everyday business and domestic users for them to be enrolled as consumers.

2 Perhaps the foremost example of such endeavours is the European Commission's Fifth Framework Programme (EC5FP), which seeks to recast its earlier approach to Research and Technology Development (RTD) within a distinctive 'integrated, problem-solving approach' (European Commission 1996).

3 Taken up, for example, by ESPRIT: the European Strategic Programme for Research on Information Technology under the European Commission's Fourth Framework Programme. The Fifth Framework Programme takes this even further, linking research and development with support for commercialisation and uptake within an integrated programme. And its activities in the field of Information Society Technologies are qualified by the phrase 'user friendly'.

4 This is not to suggest that outcomes are entirely a local achievement. Indeed, some have argued the need to take on board forms of 'soft technological determinism' – for example that the availability of certain technical facilities by changing the time, money and other forms of investment needed to achieve certain results changes the parameters on which paths of action are chosen. However, the patterning achieved by technologies is not a simple result of the functional characteristics of an artefact; its symbolic dimensions are also important. Societal outcomes of technological change are thus a complex product of the interaction between a range of players involved in the supply and use of a technology. They cannot be understood simply by snapshot assessments of the immediate settings, but must be examined as in their longer-term and broader context. These considerations suggest that conventional ideas about how to conduct socio-economic 'impact assessments' for new technologies are more or less meaningless.
5 Network technologies include ICTs but are not necessarily based on physical networks.
6 The relationship between suppliers and users may be particularly difficult given the uneven distribution between them of technical and other pertinent knowledges (for example of the application domain). This is reflected, for example, in the 'difficulties in communication' frequently experienced between technical specialist suppliers and non-expert users.
7 Software may thus be a site of conflict and controversy between diverse groups with different objectives, commitments and priorities (Dunlop and Kling 1991, Quintas 1993, McLaughlin et al. 1999).
8 That software is designed to be configured by the user does not, of course, mean that this process of adaptation is smooth and without costs; packages may not meet al.l user eventualities, and the costs of customisation may result on default values being adopted – the 'power of default' (Clausen and Koch 1999, Pollock and Cornford 2001, Pollock, Procter and Williams 2003).
9 Such considerations may be particularly significant in relation to the main ICT platform for the home – it was perceived that choices here could constrain the types of applications that the mass of users would readily be able to access, and the range of roles that these consumers will have. For example, will they mainly be passive consumers of informational products (on a broadcasting model) or will they be able to engage in more interactive uses, or will they even be information producers in their own right? This range of scenarios encompasses very different images of the character of the future 'Information Society', who will be included and how. Dichotomised stereotypes emerged. Will we all be producers of web pages? Or will it be used for dialling a pizza?
10 A similar observation could also be made about the commercial and individual consumers of ICT.

2. The Scope and Methods of the Study

GOALS OF THE SLIM PROJECT

The Social Learning in Multimedia (SLIM) research project emerged from discussions amongst an informal network involving some of the leading European research centres in the area of *social shaping of technology* – and in particular in the social shaping of Information and Communications Technologies. The consortium sought an *interdisciplinary* understanding of the *process* of technological change – addressing the technical dimensions as well as different social science traditions and which was in particular open also to insights from *cultural and media studies*, which were judged to be of special importance to analysing the development and uptake of ICT.

The overarching goal of the study was an improved understanding of the relationship between technology and society in the case of dynamic and complex technologies such as ICT. A new framework was developed (to be discussed in Chapter 3), based on the concept of 'social learning', for analysing and perhaps intervening in technological change. The main empirical part of the study, reported in Part A of this book, which examines processes of **social learning** in ICT application, uptake and use, focusing upon the ways in which current and potential requirements of different groups of (actual and potential) users are represented and matched with technical possibilities. These were addressed through a set of **integrated studies** of the development, implementation/consumption and use of new ICT-based products and services. Chapter 2 explains the research design and methodology for these integrated studies.

Our study was concerned to explore how social learning and innovation in ICT might best be focused on meeting existing and emerging user requirements, through social experiments, commercial trials and other means. We therefore also explored public policy options for the appropriation of ICT, and how they might best promote innovation and social learning.

What, then, are the implications for public policy at regional, national and European levels and for the practice of digital experiments? This is the main focus of the other part of this study, reported in Part B. Chapter 9 pulls together the policy implications of our research. It builds upon a series of national studies, undertaken by each of the research centres as part of a cross-national comparison exploring how the national/regional setting shaped the strategies for the development and appropriation of ICTs, and also addressing the influence of the broader international setting. The method and conduct of the national studies, together with their findings are discussed in Chapter 8.

DESIGN AND CONDUCT OF THE INTEGRATED STUDIES

The attempt to study both the development and implementation/use of ICT applications represented a distinctive feature of the SLIM project and perhaps its most ambitious goal. Earlier studies of everyday technologies had typically failed to do this. With few exceptions (notably du Gay et al.'s 1997 study of the Walkman) studies have ended up looking at EITHER development OR consumption/use of particular artefacts, even where they set out or claimed to do both (see, for example, Thomas and Miles 1990, Cockburn and Furst-Dilic 1994). The reasons for this are not difficult to find. The most immediate factor is the long timescales for both the development and adoption of new artefacts, which exceed the duration of most externally funded socio-economic research projects.[1] In addition, development and use may be separated in geographical and institutional terms.

We drew, wherever possible, upon existing studies conducted within the research team, to allow longitudinal insights as well as historical reconstruction of previous phases of development. In this way an array of extensive investigations was achieved, exploring the processes of development, implementation and use of ICT artefacts. The studies allowed varying insights into appropriation processes in different settings. Some studies were of projects that remained mainly at the development stage and did not provide an opportunity to address appropriation.

An array of cases was developed in relation to three areas of activity,[2] chosen because of their special significance as probable 'drivers' of the anticipated digital revolution and their social and economic importance. They were: education; cultural industries/ entertainment; public services.

At the outset of the study we conceived these studies as first addressing how the supplier conceives new ICT applications, and thereby 'constructs the user'; then assessing the extent to which these constructs match the expectations of actual and emerging users. By examining the consumption and use of new ICT offerings, we sought to explicate processes of **user innovation**, by which the user fits ICT products to their own purposes highlighting the extent to which this may simply replicate supplier concepts or may alternatively extend them in unanticipated ways. In practice the cases revealed a rather more complex and inchoate situation than might be presumed from this initial schema. For example, as we shall see, the tacit expectation that design would be informed by a more or less explicit 'representation' of the user was by no means fulfilled. More generally, the separation between an initial design and subsequent implementation/consumption phase was not always clear-cut. To aid comparison between cases which differed to a greater or lesser degree in scale, time frames and their socio-technical setting, a common analytical frame was used. This started by mapping the 'translation terrain' of the project, including the various actors directly involved in the ICT project (in relation to design, implementation and use), and the array of actors and institutions constituting the broader context for the development.[3]

Relevant cases were developed in each area of activity by the SLIM Research Centres in Belgium, Denmark, Ireland, the Netherlands, Norway, Switzerland and the UK. Three centres took on the role of the 'lead site' for each area, taking responsibility for coordinating the selection of cases, their analysis and writing up the substantive findings for that sector. A further 'cross-cutting' analysis was undertaken, pulling together themes and findings about the development and use of ICT systems within 'the organisation' and issues arising there. Though primarily drawing upon fieldwork and research undertaken for the integrated studies, it was supplemented by selected additional studies of organisational adoption.

The University of Edinburgh team took responsibility for the overarching analysis of the results as a whole and their policy implications.

These studies are summarised in the next section. A more detailed discussion is included in the analytical sections of the book (Chapters 4–7), where we also present a number of case-summaries highlighting insights from particular cases.

Education[4]

There has been considerable interest in the application of ICT in Education, though progress has been uneven and early predictions that e-learning would rapidly and profoundly transform education institutions and practice have not been realised (Collinson et al. 1996, Pollock and Cornford 2003). Perhaps the greatest progress has been made in higher education, where a strong technological infrastructure and expertise in ICT technologies/techniques, is matched by understanding of the education process and the substantive knowledge needed for the content of programmes. Conditions in other areas of education – such as primary and secondary schools, industrial training and education/'edutainment' in the home – may be rather different in terms of access to such financial and technical resources and pedagogic challenges, presenting bigger initial obstacles to the uptake of educational technologies. A particular interest was with the interaction between ICT design and choice of pedagogy. The education study explored how the adoption of ICT differs between different sectors and sites of learning (van Lieshout, Egyedi and Bijker 2001).

Cultural industries/entertainment[5]

This element of the SLIM project focused on cultural products and services which are directed at the final consumer, individual households and citizens. The development in these industries of new digital media will constitute one of the biggest commercial markets for ICT, as well as having direct social significance. Studies in this domain drew attention to the importance of the cultural content of ICT as well as its more narrowly 'technical' features, and made a special contribution to its conceptualisation.

This research examined the particular conceptions and definitions of the new 'digital media' of 'users' (and of markets) which informed the development of these innovative products and services in the relevant national and regional contexts in an era of increasing globalisation of cultural and social relations (Preston 1999, 2000)

Public administration and information[6]

This part of the study focused upon public administration and information systems. The opportunity arose for a more detailed focus upon the important European development of 'digital cities' including the original Amsterdam Digital City (DDS) (van Lieshout 2001) and a number of similar initiatives that emerged in its wake (Lobet-Maris and van Bastelaer 1999, van Bastelaer et al. 2000, Jaeger 2002).

The organisation: cross-cutting study[7]

This cross-cutting study addressed the process of implementation and appropriation of ICT configurations in organisational settings, bringing together findings from relevant integrated studies and a limited amount specific additional fieldwork in firms that have been in the forefront of adopting ICT. It explored the range of organisational applications and the extent to which the uptake of ICT varies between different organisational activities and settings.

A large number of ICT products are becoming available for use at work (e.g. desk-top video-conferencing, CD-I, the Internet, CSCW tools). Organisations have started experimenting with this bewildering array of offerings to determine which are appropriate for different kinds of organisational functions (such as training, cooperative telework, external communication). This poses new problems for the management of technological change. An interesting feature is the combination of standard tools and components, and their customisation to meet the particular the technical and organisational setting. The development of these specific ICT organisational configurations is of particular interest as an arena of social learning.

THE SLIM CASE STUDIES

The following describes the case studies undertaken by the SLIM researchers. There is not space here to present a detailed account of each case (and which would, if taken together, also be rather unreadable). The individual integrated studies have, moreover, been published elsewhere: Education (van Lieshout, Egydedi and Bijker 2001), Cultural Industries (Preston 2000 and 2001; Preston and Kerr 2001), Public Administration (van Bastelaer et al. 2000). We have therefore briefly described here what each project was intended to do and what central themes emerged from the research on that project. The case studies are presented by stream: cultural content; education; public administration; and organisations. More detailed accounts are presented within the analytical chapters as illustrations of particular social learning processes.

Cultural Content

Nerve Center – the virtual museum of Saint Colm Cille (Case Summary 6.1) *IRL*

This study describes the work involved in developing a virtual museum of Derry's Saint Colm Cille undertaken by a local community media group and the issues involved in digital presentations of this type.

It concentrates on the rearticulation of cultural content in new media, addressing the new types of narrative and navigational structure in the respective forms. It highlights the diverse skills and learning processes around the exploitation as well as the development of new digital media products (Kerr 2000, 2002).

RTE – The Den on the Net: global technology traditional media and cultural content (Case Summary 6.7) *IRL*

This case addresses the experiments with new digital media made by a traditional national public service broadcaster in an attempt to expand their existing services and develop new revenue streams.

It highlights the problems and potential for capitalising on new media for developing revenue streams and the possibility of developing new markets in the Irish Diaspora (Kerr 2000, 2002).

Compuflex – Compuflex attempts to enter the global digital media content market (Case Summary 6.2) *IRL*

The study discusses the ways in which a large multinational software company, primarily engaged in the production of task oriented software, sought to move into on-line content and service provision.

A central theme concerns how international content is localised. It focuses on the need to make content relevant to local audiences in order to ensure that it is consumable – a task which is more than simply linguistic translation. The study also focuses on issues around media convergence and their relationships to content generation (Kerr 2000, Preston and Kerr 2001, Kerr 2002).

Local Ireland (Case Summary 5.5) *IRL*

The case explores the development of a web-based service involving the provision of locally generated content and other communications services.

Focusing upon the localisation of content for consumption in Irish markets, it highlights the relations between local users and potential sponsors and the problems in developing markets for such local media (Kerr 2000, 2002).

Black Out – integrating interactivity into narrative forms (Case Summary 5.3) *DK*

Centring on innovative CD-ROM development in Denmark, this case study examines the social construction of new digital media forms.

It shows some of the problems and potentialities of developing new CD-ROM products and the importance of some idea of audience in their development, and contrasts the technical potential for radically restructuring narrative forms with predominantly conservative forms adopted in most products (Hansen 1998).

Girls ROM – constructing content around gendered cultural forms (Case Summary 5.4) *NO*

This case concerns a CD-ROM ('Girls-ROM' or Girls Room) given away with the magazine *Det Nye* aimed at young women, and subsequent developments in the magazine industry aimed at young women and their use of ICTs.

Various articulations between gender and computing are possible, and the case focuses on social learning around this and the construction of (gendered) markets.

Girls-ROM oriented itself towards existing cultural forms (e.g. an electronic diary) and delivery systems (PCs with CD drive) in order to open up the gendering of ICT markets. The magazine publisher subsequently launched a service (HomeNet) which addresses women as a specific market segment within a broader service oriented to different (gender, age, etc.) categories of users. This may indicate growing awareness by mainstream media corporations and advertisers that it is in their interests to address 'exclusion' and construct the widest possible audience/user base (Spilker and Sørensen 2000, 2002)

Education

Telepoly – matching learning technologies and teaching strategies (Case Summary 4.2) ***CH***

This study examines the strengths and weaknesses of a broadband teleteaching initiative and the social learning that took place between a number of actors in the project and the eventual outcome of these interactions and their effects on the service.

It shows how social learning can contribute to an analysis of the development of complex ICT systems and their scope, and gives an indication of the negotiated character of such developments (Buser and Rossel 2001).

Cable School – technical infrastructure at the expense of the social dimension (Case Summary 4.3) ***UK***

This study examines an attempt led by a cable television network to connect Edinburgh and East Lothian schools to the Internet, and the organisational implications of such a project. Cable Co. pursued this as a way of projecting its capabilities for providing ICT infrastructure to local authorities.

The system did not meet its objectives, and only some elements survived after support ended. This highlights the importance of incentives and support for the development of expertise in schools to support the uptake of ICTs in education (Slack 2000a).

Teleteaching in Bornholm – learning new means of interacting for new media (Case Summary 6.5) ***DK***

This case examines the development of distance education by video-conferencing on the island of Bornholm and the problems encountered in the development of the pilot, especially by teachers and students in the use of an apparently less than flexible system.

It shows the ways in which technology configures both learners and learning, and how teachers and students cope with this – especially in terms of the need for novel ensembles of technologies such as email, video-conferencing and so on. The study highlights the complex and lengthy problem-solving processes involved in getting technology to work and in developing forms of teaching and interaction to support the potential afforded by the technology (Hansen and Clausen 1998).

Language Course – technology and education as marketing strategy (Case Summary 7.2) ***NL***

Developed by the marketing and communications division of an international electronics and communications services supplier, this project offered language education and was displayed at an exhibition of the company's products.

The study shows how the exigencies of marketing shaped the development and usability of teleteaching technologies and materials, and how teachers and students evolved coping strategies for these contingencies. It notes how students took advantage of features of the technology in classroom deviance in devising playful uses and subsequent design modifications to allow teachers to retain control! It also details how far success and failure in demonstrations are socially constructed and the impact this has on technologies both in terms of marketing and use/uptake (Mourik 2001).

Spice Girls to Cyber Girls – reconfiguring the gendering of computers (Case Summary 6.3) ***NO***

This study of an emerging sub-culture of computer-enthusiast young women explores the relationships between female students and computers, showing that female students tend to see computers as a 'play pal' rather than a tool. Starting with traditional home-based games, these girls come to use the computer for intimate social communicative roles, like writing a diary and poems, and as a communicative device, as a gateway to Internet. The study points to the importance of play with technology and argues that play activities around the computer should be supported in training and education as a means of enfranchising female students in ICT use (Nordli 2001).

Coffee and Computers? Cybercafes – a site for the local appropriation of ICTs (Case Summary 6.4) ***UK***

This study of a number of cybercafes in Scotland and Ireland points to the importance of 'appropriation intermediaries' – cybercafe managers and local experts – in coupling generic ICT offerings to local groups of users – and the different ways in which these generic capabilities and spaces may be configured (Stewart 1999, Laegran and Stewart 2003).

Teaching Transformed – ICTs in Norwegian education policy ***NO*** **(Case Summary 9.1)**

This study discusses the development of ICT policy in Norway's education system and the role of social learning in the provision of computing technology within the classroom. It highlights shortcomings in the complex 'learning economy' around ICTs in education. Weak links between teachers and with managers mean that social learning is poorly exploited and has to be repeated in different local settings. Top-down policy interventions need to be integrated with local initiatives to support classroom ICTs to kindle teacher enthusiasm and integrate ICT in teaching (Aune and Sørensen 2001).

Hypermedia Use in Education – a diversification Experiment (Case Summary 7.1) NL

This study examines the development of a project to use ICTs in universities in the Netherlands, in an experimental application of hypermedia in a new visual culture course.

It shows how an ICT teaching project diverges in the course of its development, highlighting the constraints experienced by the students using the technologies; the ways they find of working around these and the emergence of unexpected uses. Assisted by the efforts of teachers to resist 'closure' about the implications of hypermedia helps this project function as a 'diversification experiment' (Egyedi 2001).

Public administration and information (digital cities)

Geneva MAN ***CH***

This study of a pilot project to develop high-speed links within the city of Geneva shows how technology can be used in urban contexts and highlights the importance of users in design. It further comments on the relationship between deregulation of telecom markets and the role of PTTs in the development of large-scale projects.

Craigmillar Community Information Service (CCIS) – learning to be fundable (Case Summary 4.4) UK

This community telematics initiative in Edinburgh to connect a deprived community to ICTs and to regenerate the area through ICT use and ICT literacy highlights the importance of social learning along two dimensions:

First, in understanding the development of services as contested terrains – both in terms of the idea of a community service, the perceived community of users and its relationship to the service. We analysed the competing constructions of space and place around on-line services geared towards identity and locality (Slack and Williams 2000).

Second, we analysed, under the rubric of the social learning of independence, the creative role of key players in intermediating between funding bodies, various groups from the local community, and other potential users in a constantly changing technical environment.

Copenhagen Base (Case Summary 5.1) ***DK***

The study of the Copenhagen Base (CB) is a story about the development of a database of city information, which was initially conceived as a tool for staff in the administration. It was set up as an information kiosk and was based on the city's internal computer system. After a short period, it became clear that the user group should be expanded to include the citizens of the city. Thus CB was transformed into a web-based service and launched on the Internet. Through various means this project developed a user-centred interface and involved users both among the civil servants and the citizens in the design and population of the service.

The case study highlights the potential disjunctures between technical and 'lay' versions of the service and draws attention to the need to include citizen input to assist the development and uptake of a usable service (Jaeger 1999, 2002).

Amsterdam Digital City (*DDS*) NL
DDS – an evolutionary model? (Case Summary 4.1)
The Interface Design for Amsterdam Digital City (Case Summary 5.2)

Usually regarded as one of the first in Europe, Amsterdam Digital City, this study examined the development and use as well as the rearticulation of the digital city over time. It exemplifies an evolutionary model for system development, as the boundaries of those involved, conceptions of the user and the design of interfaces change repeatedly as DDS develops and grows.

The study flags the importance of metaphors that can link, for example, designers and users, and the development of a digital city as a role model for others. It stresses the importance of partnerships between various groups (e.g. media and technology specialists, community groups) in social learning and the need for autonomy and experiment in design and building (van Lieshout 2001).

DMA: Antwerp Digital Metropolis *BE*

A partner of DDS, the study focuses on the integration of services in ICT as well as the need to enrol user constituencies through novel technologies such as web booths. It draws attention to the importance of integration and presentation of citizen requirements and stresses the need to make web services available to all in innovative ways.

Pericles *BE*

The study details the technology-driven Pericles project that was developed in the French speaking part of Belgium and the interactions between users and administrators within the project.

It points to some of the problems arising from the technology push approach, and shows how the lack of clarity in an overall plan for a project can impact on development per se, and suggests some insights from social learning as to how this might be remedied.

Frihus 2000 NO

A development of information superhighways for rural areas, this case focuses on user involvement and technological capabilities over time and their effects on the service.

The study stresses the importance of democracy and user involvement, especially in the generation of requirements – even where these compete with those of technical specialists.

Organisations

Edipresse *CH*

This organisation adopted new technologies as a way of coping with competitive challenges in the publishing industry – the study highlights the limits of tailoring packages and the development of working practices after the introduction of ICTs.

It stresses the social learning aspects of tailoring packages within an organisational framework and the organisational implications of ICT introduction.

UK Bank – experimenting with desk-top video-conferencing (Case Summary 6.6) UK

This organisation conducted a series of pilots in different areas of banking business to assess the utility of desk-top video-conferencing (DTVC) to the organisation. Standard packaged components were knitted together and cut down to provide a limited set of functions (phone and video connection, joint access to application software and data).

In this way the organisation was able to learn about the affordances and constraints of the technology, and the kinds of interaction for which DTVC was suitable. Individuals developed ways of working that best exploited these technical capabilities (Procter, Williams and Cashin 1999).

NOTES

1 Indeed, this is one of the reasons why those studies which encapsulate development use of a technology are almost without exception, retrospective historical accounts.
2 This focus on areas of human activity, rather than existing industrial sectors, was sought because of the potential of ICT to cut across existing institutional boundaries and locales.
3 Charting this translation terrain in diagrammatic terms proved an effective tool in communicating the features of case and drawing attention to differences and similarities that might otherwise have been overlooked.
4 Lead site: the University of Limburg.
5 Lead site: Dublin City University.

6 Lead site: Namur University (FUNdP).
7 Lead site: Lausanne EPFL.

3. What Do We Mean by Social Learning?

DEFINITION AND MOTIVATION OF SOCIAL LEARNING

From social shaping to social learning

This study has drawn upon the concept of 'social learning' to capture the complex dynamics of innovation in advanced applications of ICT. In this chapter we *define social learning* and explore how, in the course of this research project, we have developed our conceptualisation of social learning to achieve a more detailed and nuanced understanding of the process of technological innovation and address the many actors and spaces involved. The social learning framework we are developing here is not wholly distinct from a social shaping of technology (SST) perspective – but tries instead to carry forwards some aspects of SST and address earlier weaknesses.

Our intellectual project was to further enhance 'social shaping' studies by promoting a dialogue with other strands of research; notably:

1. *Work from a cultural studies perspective and on the symbolic dimensions of artefacts.* Early SST studies have been criticised for their emphasis on the initial design phase at the expense of the consumption of technology and for their neglect of the symbolic dimensions particularly around consumption (Mackay and Gillespie 1992). Another area of work which has been largely overlooked by the more rigorously established social sciences is studies of consumption from a marketing perspective (see, for example, Mick and Fournier 1998). In developing the SLIM project we deliberately sought to address this by including in the research consortium researchers and research centres from cultural studies and semiotic perspectives whose work focused upon the symbolic dimensions and the 'appropriation' of technology.
2. *Related evolutionary models of technological development.*
3. *Work from organisational studies on 'organisational learning,* (for example Schon, 1983) and related work which has emphasised the adaptive and reflexive capabilities of actors. In relation to this last point, our use of the terminology of social learning rather than SST also signals a focus not only upon analysing processes of innovation but also upon the *possibilities for intervention* – and in particular on the scope

for adjusting innovation processes and their outcomes as a result of the *reflexive activities* either of the players involved or of technology researchers (Wynne 1995).

In short, our interest in social learning represents an attempt to broaden the intellectual base of SST and to broaden and clarify opportunities for intervention. We sought to explore these concerns in particular domains of 'social learning' around the application of ICT.

Social learning – developing an appropriation perspective

The social learning project thus focuses upon understanding how generic technological capabilities get matched to, applied within and eventually form part of diverse social settings and activities. In short, social learning seeks to understand how such technological capabilities are appropriated and become embedded in society.

In developing this perspective we are seeking to overcome the privileging of technology design and development that has prevailed in much discussion to date. The rhetorics of technology supply, and much early work in technology studies, conceptualise design as readily meeting the purposes of designers and the future users they must cater for, and conversely downplay the contribution to successful innovation of a wide array of other players – the intermediate and final-users of technologies, and others indirectly involved. We advance a critique of this 'design fallacy' (see Chapters 4 and 5) and related technology supply-centred perspectives, arguing that the analysis of design must be located within its broader context of multiple cycles of technology design–implementation–use. We lay out an alternative perspective that pays attention to the appropriation of technology, drawing upon research into technology implementation, consumption and use. The extension of the focus of enquiry to include technology appropriation and use has important implications for how we theorise technology and for intervention in policy and practice. It widens our view of the locales of innovation, the range of players involved and their respective contributions, and highlights, for example, the *diversity* of users (and their specific expectations, contexts etc.), and their *active role* in developing practices, concepts of use and meaning around artefacts (Russell and Williams 2002a). The appropriation-centred view is a necessary corrective to the traditional privileging of technology design. The social learning perspective incorporates this and further addresses the interaction between technology development and its appropriation.

Definition of social learning

Social learning, thus conceived, seeks to provide an inclusive analytical framework. For example, as outlined above, it encompasses cycles of development and consumption of technology as well as the bringing into being of the broader circumstances needed for the technology, including the formation of wider technology regimes.

We use the concept of social learning to provide tools for addressing the broad spaces in which technological artefacts and systems are appropriated and made part of (or rejected by) particular social systems and settings (Rip, Misa and Schot 1995), and how they may themselves be transformed in the process. We should more accurately have referred above to 'socio-technical systems' here since social learning offers a profoundly socio-technical perspective, emphasising the intimate interplay between humans and artefacts across many settings.

Fleck (1988) argued that the 'implementation arena' was a laboratory for the further innovation of industrial artefacts emerging from technology supply. The concept of social learning takes this further and suggests that we should see wider 'society' as a laboratory for learning about generic technical capabilities (Herbold 1995).

Put this way, the term 'social learning' sounds evidently desirable; however, we do not imply that the processes are smooth, homogeneous and consensual. We express caution regarding the interpretation of the term 'learning'. Social learning is not for us a narrowly cognitive process (we are thus using the term 'social learning' in a very different way to its educational usage, or in social psychology). We are seeking instead to understand processes of socio-technical change, as also a process of negotiation and interaction between different players and thus subject to conflicts, and differences of power and interest. (For a more detailed discussion see Sørensen [1996], on whose original exegesis, this section draws heavily.)

We explore below the intellectual debt to the ideas emerging from economic history and evolutionary economics of 'learning by doing' and of 'supplier–user interaction'. However, social learning is much more than this. As Sørensen (1996) puts it:

> Social learning can be characterised as a combined act of discovery and analysis, of understanding and giving meaning, and of tinkering and the development of routines. In order to make an artefact work, it has to be placed, spatially, temporally, and conceptually. It has to be fitted into the existing, heterogeneous networks of machines, systems, routines, and culture. (Sørensen 1996: 6)

Thus, to analyse social learning we have to go beyond the narrow instrumental understandings of economists and address issues of meaning and identity as well (Sørensen 1994a). The importance of this symbolic effort – to make technology meaningful – is of obvious relevance when we come to look at domestication and the cultural appropriation of novel technologies. The need to address the symbolic as well as functional dimensions of artefacts is particularly relevant with information and communication technologies: to the extent that they function as *media* that carry various kinds of information and cultural content. This first reminds us of the need to not just restrict our inquiry to hardware and the material nature of things, but to focus more explicitly on the symbolic dimension. Second, and more specifically, ICT artefacts can be seen to have a 'double symbolic' dimension (Preston 2000): not only does the physical artefact have meaning; there is also the meaning that the information and cultural content conveys, both through the meaning of this content and the particular form in which it is presented.

Similarly, social learning is not restricted to the operation of the 'learning economy' of supplier–user interactions, but extends into the efforts of various players associated with ICT to regulate and provide order in and reshape the broader context. We have characterised this as learning by regulating. It involves both of the main types of social learning we have identified: *learning by doing* in the 'trial and error' process of developing their own strategies and orientations around technology, and *learning by interacting* in their efforts to align and regulate the behaviour of other players.

As we have noted, our approach draws upon a number of intellectual traditions including economics, organisation studies (e.g. Schon's 1983 work on 'organisational learning', culture and consumption studies. We are seeking to develop an analytical perspective that pulls together elements from approaches which have hitherto been not well integrated (for example technology development and consumption; its cultural and its material aspects). We are, moreover, trying to do this in a context in which the particular terms have been used in often inconsistent ways. We therefore start our analysis by exploring the origins and definition of these different elements.

ORIGINS OF THE CONCEPT OF SOCIAL LEARNING:

Learning by doing

Economists and economic historians became interested in social learning through studies of productivity that showed continuous improvements in performance of equipment over long periods of time without any investments in new technology. Arrow (1962) calls this phenomenon 'learning by doing'.

The idea is that workers, individually as well as collectively, develop more efficient ways of employing machinery through their experience from usage. This kind of 'learning curve' effect is well-known. A related phenomenon is 'learning by using'. Rosenberg (1982) suggests this concept to describe the process through which a user gains familiarity with a given piece of technology and develops skills in making use of it.

These concepts point to the fact that the properties of a technology (its utility; its affordances; its limitations) may not be immediately apparent, but are discovered and learnt through experience, often in relation to particular productive processes and activities.

Such learning by doing provides a potentially very important source of information on the effective use of a technology. By giving suppliers access to what users have learnt about their products and what deficiencies and potentialities they have discovered, it could provide invaluable information for subsequent product innovation. It has been further noted that this information is often not systematically collected and used – perhaps because of the strength of the rhetorics of technology supply (which effectively suggest that, if a new product already fulfils user requirements as claimed, what need is there to examine the problems that may arise in its implementation and use). This underlines the importance of the *linkages* between users and producers that can act as a vehicle for this kind of knowledge exchange. To innovate successfully, producers may depend critically on information from users, and vice versa. This is the basis of the idea of learning by interacting and the learning economy (Andersen and Lundvall 1988).

Learning by interacting – the learning economy

The idea of learning by interacting has emerged importantly amongst evolutionary approaches. Evolutionary models of development of technology have a long history in technology studies (see, for example, Gilfillan 1970). They sought to understand the complex interaction between technology supply and use by applying evolutionary metaphors of the generation of variation and selection (see, for example, Schneider 2000).

Freeman's classic work (Freeman 1974) highlighted the importance of supplier–user coupling for successful innovation. Some stability in inter-firm relations, as well as a supportive institutional environment, may be needed to provide the necessary preconditions to maintain such linkages in the supply chain. Evolutionary accounts have analysed these linkages as constituting a learning economy. In seeking to understand differences in the regional and sectoral performance of economies, Andersen and Lundvall (1988) emphasised the institutionalisation of such linkages through the development of channels of communication, shared codes of conduct and conceptualisations between suppliers and users. The operation of the learning economy is shaped by the systemic institutional features and setting of a given regional or national economy, including public policy.

The concept of the learning economy calls into question simplistic beliefs in the market as a mechanism of communication as it demands a greater stability of economic relations and more developed patterns of communication than those held up by idealised market forces alone (Sørensen 1996). There may also be a need for new institutional arrangements in which governments – locally, nationally and even supranationally – may play a crucial role.

Supplier–user communication can be relatively straightforward in relation to industrial technologies, where the high value of transactions may motivate collaboration and there are direct links between supplier and consumer, and where both supplier and consumer share relatively high levels of technical skills as well as presumptions about the values associated with technology. Supplier–user relations, conversely, can be expected to be more problematic in the case of mass-produced consumer goods, where there is little scope for direct engagement between suppliers and the dispersed mass of anonymous users who are only indirectly linked to the supplier through the market. To return to Hirschman's (1970) frame of reference, the consumer has – in particular in mass markets – only a choice between exit and loyalty, and thus very limited possibilities of 'voice' – communication with producers/ designers. Particularly in relation to novel products where there is no established customer base (where, indeed, the customers do not even exist yet), the supplier may have little understanding of the user and the use setting. The supplier is thus forced to play a major role in prefiguring, or indeed constructing, the customer and the market, and must find some way of modelling user requirements. Fortunately, many companies recognise this and make efforts to reproduce user–producer relations, at least on a semi-permanent basis, for example, by conducting market research on 'proxy users' (e.g. panels of potential customers) or through discussions with intermediate users (e.g. retailers of the goods and services being produced) (Akrich 1995). It may also be necessary to enrol a range of other actors, such as competing suppliers and suppliers of complementary products (Collinson 1993). This remains a rather difficult and uncertain process for many suppliers – particularly in relation to the mass market for technologies for everyday life. Thus in relation to domestic technologies problems may arise because the home remains a largely private sphere, with only weak, and predominantly indirect linkages between supplier and user (Morley and Silverstone 1999, Silverstone 1991, Silverstone and Hirsch 1992, Cockburn and Furst-Dilic 1994, Cawson, Haddon and Miles 1995).

Evolutionary economists have been concerned to characterise and explain the differential 'revealed performance' and outcomes of these national or regional innovation systems (e.g. Andersen and Lundvall 1988). What is more interesting for our purposes, however, is to understand the detailed process of innovation: how learning by interacting takes place in relation to particular technologies and in relation to particular groups of actors. Addressing the substance of these interactions in detail offers important insights. For example, much of the knowledge that emerges from local efforts (learning by doing) is experience based, contingent and tacit (Williams, Faulkner and Fleck 1998), and cannot simply be captured, transported and applied in other contexts. It is not a linear process of knowledge transfer. Instead, its wider application will depend upon a creative process of knowledge translation: selecting relevant experiences and transforming them. Here the social learning perspective goes beyond the economists' focus on linking structures (whose operation is demonstrated by the revealed cumulative performance of players in the innovation system) to address the detailed mechanisms and substance of this learning by interacting.

Innofusion

Scholars within the field of Social Shaping of Technology thus sought to draw different insights from these same coupling processes analysed by evolutionary economists – through detailed studies of technological innovation. Thus Fleck's (1988) concept of innofusion (= innovation in technology diffusion) contested traditional views that innovation was restricted to research and development, and stopped when a product left the laboratory. Instead, he argued innovation continued as these artefacts were implemented and used. He drew attention to the ways in which supplier offerings were transformed and further innovated as they were implemented and used in the struggle to get the technology to work in useful ways at the point of application. He highlighted the possibility of feedback from implementation to future technological supply – a possibility which depends critically upon kinds of linkages and channels of communication between supplier and user which make up the learning economy.

The concept of innofusion provides an important tool for analysing social learning processes in the adaptation and transformation in use of technical artefacts. However, it was developed to account for innovation in industrial technologies. It does not per se, provide an adequate framework for analysing the ways in which artefacts become incorporated within practices and acquire their meanings and significance. To address these dimensions we have turned to work from cultural studies and to research into the consumption of technologies especially in the settings of everyday life.

THE APPROPRIATION FRAMEWORK

We are seeking to develop our framework to analyse social learning processes around the application, consumption and use of technologies. Our aim is to be able to analyse in greater detail the nature of 'learning by using'. What is the scope of such learning, what is achieved, how does it affect the outcomes of technology adoption and how may such insights be helpful to the design of better technology?

Here we draw from anthropological or ethnographic perspectives the concept of 'appropriation of technology' (Pacey 1983, du Gay et al. 1997). The concept of appropriation reminds us that the adoption, consumption and use of artefacts is an active process (in contrast to the passivity implied by traditional terminologies such as diffusion); it highlights the degrees of freedom remaining to users, and other players outside technology supply, in exercising choices around the selection and local deployment of artefacts, and, finally, it emphasises the symbolic as well as instrumental aspects of these processes.[1] In particular, we emphasise that technology has to become encultured (or embedded) in order to function (Lie and Sørensen 1996, McCracken 1988, Silverstone and Hirsch 1992).

Similarly, media and consumption studies focus on the symbolic aspects of technology and the need to analyse how artefacts acquire their meanings. For example, McCracken (1988) draws attention to the many rituals which people employ to give meaning to material objects. Without meaning, artefacts remain artificial and alien. Work has to be performed to integrate a technology into a culture.

This is important for our understanding above all, in that it emphasises that social learning involves the creation of meaning as well as practical efforts to make technology work. The interpretation of an innovation is of great importance to its eventual success or failure. The reinterpretation of the camera from a professional to an amateur artefact (Jenkins 1975) or of the video cassette recorder (VCR) from a tool for broadcasting companies to a household appliance (Roosenbloom and Cusumano 1987), extended in a dramatic fashion the market for these products. The future role of ICT may depend on the public understanding of this technology in terms of moral acceptability (Is it harmful to children? Is it mainly 'toys for the boys'? etc.).

Meaning is projected internally as well as externally. The objects we use should be compatible with our personal identity as well as with the perception of self that we want to convey to an outside world. This is one of the reasons why some technologies are met with greater enthusiasm from some groups of people than other. Reactions towards new technologies that are often labelled 'resistance', may on closer examination prove to be rooted in this dynamic of meaning: the common interpretation of the innovation is not in accordance with our self-identity.

Decisions about appropriation may thus depend on a sense that a technology is socially appropriate – in the sense of proper.

The definitions of the word 'appropriate' (a.n.) in the *Oxford English Dictionary* chart the parallel evolution of the term as:

2. Belonging to oneself; private; selfish. (1627)

3. Assigned to a particular person; special, individual (1796)

5. Specially fitted or suitable, proper (1546)

Similar sets of definitions with correlations of meaning between concepts of ownership, uniqueness and fitness can be identified in a range of closely related terms: property, propriety, proper.

Despite this interest in the appropriation of technology, in our analysis we have resorted to the related concept of 'domestication', for several reasons: There is no general theory of appropriation. No dominant definition or methodology for analysing it has yet emerged – or seems in prospect. The term 'appropriation' has been used in rather different ways by authors from various disciplinary backgrounds and purposes. It can, for example, be applied at different levels, ranging from individual engagement with an artefact to how a country reacts to the availability of a set of global technologies. At the level of particular artefacts, the term 'appropriation' has been used to refer to both:

1. The ways in which generic technological capabilities are applied and incorporated into products and services intended for particular social settings (the home, the workplace, particular regions and nations etc.).
2. The related processes by which these products and services are incorporated into particular uses in these settings (e.g. du Guy et al. 1997).

If needing to be specific, we could describe 1 as 'intermediate appropriation', in so far as it concerns what industrial economists would describe as intermediate consumers, and 2 as 'final appropriation', though this is perhaps the default meaning when the term 'appropriation' used in cultural and technology studies.

Domestication

We therefore turn instead to the concept of domestication to provide a more well-defined approach to the study of appropriation of technology. We resort to the domestication concept as it provides some interesting analytic elements which we feel able to build upon. At the same time it has less 'intellectual baggage' than the more established terminology of appropriation, for which the multiplicity of definitions can impede clear communication. For ease of exegesis we adopt a simple schema here which sees innofusion and domestication as both being important processes in the appropriation of technological artefacts, highlighting respectively innovation around the artefact and around culture and practices. Our perspective would, however, argue against drawing artificial boundaries between the material and symbolic aspects of artefacts: innofusion and 'domestication' cannot be separated, but should rather be seen as different facets (faces?) of a unitary social learning process. Usages of the term 'domestication' (just as appropriation) are various, geared towards different purposes and contexts around which the term is deployed[2] – and we are not arguing for or expecting a harmonisation of usage.

The concept of domestication emerged from a series of important studies which sought to understand the appropriation of artefacts in the specific social setting of the home. Silverstone, Hirsch and Morley (1992) use this term to describe how artefacts are integrated into what they call 'the moral economy of the household'. They highlight the interaction between economic and symbolic transactions within the household, and between the household and the outside world. Moral economies are negotiated spaces. To domesticate an artefact is to negotiate its meaning and practice.

The domestication concept is not referring specifically to household contexts. Lie and Sørensen (1996) define domestication as a kind of taming process: 'Metaphorically speaking, we tame the technologies that surround us in everyday life.' Domestication shows the need to 'tame' the facts and artefacts that are taken from a 'wild' outside world. Thus, domestication is a way of theorising what the cultural appropriation of technology is all about.

Basically, domestication is necessary both to make artefacts work and to make sense. Both action and meaning are important. Artefacts have to be:

- Acquired – either bought or in some other way made accessible.
- Placed – which means that they are situated in a physical, symbolic and mental space.
- Interpreted – to be given meaning within the household or a similar local context of identity as well as symbolic value to the outside world.
- Integrated – into social practices of action (Silverstone, Hirsch and Morley 1992, Lie and Sørensen 1996, Sørensen, Aune and Hatling, 1996).

Strategies of domestication thus involve practical, symbolic, and cognitive dimensions. The practical dimension of domestication refers to patterns of usage and how an artefact will be employed. Symbolic efforts are about production of meaning and the relationship between meaning, identity, and the public presentation of self. The cognitive dimension is related to learning about the artefact or the intellectual appropriation of new knowledge.

The concept of domestication is also of value because it reminds us of some of the trivial challenges that have to be solved in order to make things work. In particular, it points to the necessity of constructing routines – including working out the tasks that have to be performed and the skills necessary to do these tasks, and an appropriate division of labour – for the smooth and successful implementation and operation of an artefact (Suchman 1987, see also Nelson and Winter 1982). It is through domestication that artefacts become 'invisible' taken for granted elements of everyday life. This does not mean that these routines are inscribed in the artefact, just waiting to be discovered. People construct different routines and different meanings from the same artefact. Of course, these processes of construction are contingent upon, and may be affected by, class, ethnicity, gender and age. Studies (for example of the ostensibly straightforward task of making use of a PC [Aune 1996, Sørensen, Aune and Hatling 1996]) have emphasised the diversity of learning strategies as well as formal training, noting the importance of practical activities of *experimentation* pointing to the identity between using and learning, and *tinkering* where you learn in a pragmatic manner what you need to know in order to perform necessary tasks. The ability to do this type of work is accumulated as people develop a repertoire of adoption and use situations and develop *scripts* to handle adoption of innovations (Hirschman, 1980). Learning by trial –and error is of general importance but, in addition, users get information and knowledge from each other and through a host of different channels. The learning economy observable on the macro level has its parallel in local learning communities.

However, social learning is not just the acquisition of skills and knowledge. When we emphasise domestication, we highlight the construction and reconstruction of culture as old and new combinations of artefacts, skills, knowledge and social relations. This means that users are no longer seen as passive consumers, fed by industry. In line with the idea of 'active audiences' from media studies (e.g. Silverstone 1994, Sørensen 1994), users are perceived as struggling with the socio-material relationships that they encounter in their everyday life, being creative in solving problems, changing these relations and integrating technology into their lives (Hebdidge, 1979; Hirschman, 1980; De Certeau, 1984; Fiske, 1989). Domestication is a way of conceptualising this struggle.

To sum up, we use the term 'domestication' specifically to refer to the way in which technical capabilities are explored, meanings attributed and practices developed as artefacts are integrated within local social settings. Innofusion and domestication processes are thus crucial to the successful uptake of new ICT products and service – it is here that presumptions about the use and utility embedded in a designed artefact are tested, in which user acceptance or rejection of new offerings takes place and in which the uses and significance of technologies are finally determined.

So far we have focused upon the activities of actors directly involved in technology supply and consumption. We have distinguished two arenas of social learning involved in designing a technology and getting it to work and to be useful in particular settings:

1. *Representation of the user in design*
   When a new supplier offering is developed, the designer/developer must construct it around some kind of representation of the use and user of the artefact (though this may be largely implicit). Here the designer is trying to appropriate the future user. This is the main focus of Chapter 5.
2. *Appropriation of the designed artefact: innofusion-domestication*
   the active and creative process in which 'users' selectively incorporate supplier offerings and adapt them to their specific needs, routines and practices. Here supplier hypotheses about the use and utility of their offerings are tested and the meanings and significance of an artefact are finally established. This is the main focus of Chapter 6.

There are, however, other important kinds of social learning. Successful technologies are those which succeed in establishing a supportive socio-technical setting (Sørensen 1994a). This reminds us that social learning is not restricted to the operation of this 'learning economy', but also involves creating the broader circumstances for a technology to be successful (Sørensen 1996). This points to a third dimension of social learning. We address this through the concept of *learning by regulating*. This encompasses the activities of public policymakers, as well as promoters and other players in civil society in setting the 'rules of the game' for innovation.

TECHNOLOGY REGIMES AND LEARNING BY REGULATING

We have drawn here upon the concept of 'technology regimes' developed by Rip and co-workers, to explain how regimes become entrenched (and may favour some kinds of artefact and outcome over others), and how new regimes can seek to become established (Rip 1995, Kemp, Schot and Hoogma 1998, Rip and Schot 1999). The concept of regimes also seeks to capture the expectations, visions and sets of rules, the policies and regulations that help to shape the ways in which new technologies are developed and applied. These rules thus provide a broader framework and resources for local design/appropriation activities. Finally, this 'learning by regulating' (Sørensen 1996) also includes the attempts by policymakers to find mechanisms of influence and control that are appropriate to this evolving socio-technical terrain.

The emergence of new ICT systems, infrastructures and digital media, with apparently rather different socio-technical characteristics to its predecessors, potentially throws up an array of issues for the state. The issues include (but are not restricted to) explicit types of regulation about legal and policy issues, such as: intellectual property and its enforcement (i.e. through charging systems); privacy and data security; decency (for example in relation to pornography), fraud and the problems of policing cyber-society. In relation to ICT, and digital media in particular, the thesis of convergence has immediate policy implications, as it suggests the coming together of the fields of computing, telecommunications and broadcasting which have been subject to very different regulatory regimes and traditions. The state is also involved as a provider or guarantor of the infrastructures that may be critical to the operation of a technology. Just as the widespread uptake of the car rested upon the establishment of a host of different institutions to regulate and promote this technology, today the emerging Information society is seen to require the establishment and maintenance of an information infrastructure (Kahin and Wilson 1996).

Public policy has played an important role. As we will see in Chapter 8, there have been important attempts to learn from other countries. A fear of being left behind in technological and economic competition has been reflected in powerful rhetorics of technological imperative. Mimicry of initiatives elsewhere, mutual reinforcement and efforts to harmonise policy and regulatory requirements have all tended to encourage a convergence of public perspectives. However, at the same time the practical concerns of policymakers have also stimulated them to match policy to particular local conditions, traditions and concerns.

Private actors also make efforts at regulation, for example, by developing and trying to impose standards, producing infrastructure and – more generally – by trying to enrol political constituencies as well as customers and clients to their visions and designs.

In short, regulation, in this sense, is not just located in government institutions. We see efforts to impose order and set the rules of the game across a range of bodies and in different activities. Thus private players in promoting visions of technology futures seek to establish, and put their offerings at the heart of, emerging technology regimes. Thus today we can see suppliers simultaneously competing to pursue their particular views and interests while at the same time acting in concert to establish ideas of the imperative of the widespread adoption ICTs as we move to an information society (which we have described as the rhetorics of technology supply).

There is equally an interplay between the state's regulatory and technology promotion activities. Regulation, and in particular the formation of technological regimes, is not restricted to a 'top-level' political domain, but permeates and ultimately becomes rooted in the dispersed activities of technology design, implementation and use.

We can thus distinguish (at least) four broad types of learning by regulating:

1. First (and conceptually precedent) is the creation of mobilising metaphors and visions which inspire the development of new technological systems;
2. Second is closer to the more conventional use of the term 'regulation' in the sense of setting the rules of the game, including the activities of public regulators, (including legislators, promoters and other policymakers) and other players (including pundits, and the activities of major industry and technology players). Together these may constitute particular 'technology regimes' (Rip 1995);
3. The latter merges into the third activity – comprising more formal domains of state policy formation;
4. Crucial to the idea of learning by regulating is the attention it pays to local processes of rule-setting, as well as the 'top-down' regulatory activities of the state and major players.

We discuss the first of these below, drawing some insights from our empirical studies (in particular the set of 'National Studies' conducted under the SLIM project, which also threw light upon social learning policy formation and the role of the state in the digital experiments studied, which are discussed in Chapter 8.[3]

Mobilising metaphors for technology regimes

When considering the 'mobilising metaphors' and visions which inspire the development of new technological systems we can distinguish two levels:

1. How generic *metaphors* have been mobilised in support of broad visions of technical and social change – for example, of the emerging Information Society.
2. How particular concepts of future technologies and their applications – *poles of attraction* – may be advanced.

Over the last decade we have seen the articulation of powerful metaphors in the field of ICT – including, for example such concepts as The information society; information superhighways/information infrastructures; the New Economy; the Knowledge Society. These serve to organise, give direction, mobilise resources and set norms for programmes of technological change. Here we can see some gradual changes in terminology over time – perhaps linked to the need continually to draw attention to novelty and convey urgency – as well as elements of continuity (for example ideas that technological advance will drive rapid economic and social change, and that there is therefore an economic imperative to be at the forefront of the imputed technological revolution). As we will demonstrate in Chapter 8, public policies provide an important impetus in these developments. For example, all the industrialised countries and many industrialising ones have adopted Information society policies, and there seems to be considerable mimicry and reinforcement between these visions (Dutton 1996, West 1996, Williams 1999). There is, however, a close interaction between public and private players – with technology suppliers, severally or in combination, projecting their visions of the particular technological routes to be adopted and the kinds of economic and social benefits that can be expected.

In a changing and uncertain technological context, such visions hold out the prospect of simplifying the choices for intermediate and final-users, and others, by aligning the strategies of a range of suppliers of complementary products

We can see the deployment of metaphors as operating at a number of levels in shaping expectations around ICT:

1. Overarching views and policy pronouncements about the general character and social implications of ICT (for example, as already noted in terms of how the idea of an Information society has been embraced by public policy in very different countries ranging from the USA to Malaysia).
2. At the level of more particular concepts of technology application and use, which we find in two, related, contexts:
   a) the promulgation of particular best practice exemplars – for example the promotion of the idea of digital cities as paradigm of public information systems;
   b) the attempts by commercial players and promoters to align expectations of potential consumers, competitors and others in the market about the future configuration of ICTs. We have advanced the concept of 'poles of attraction' to explore this (Stewart 1999a). As this is a novel concept that arose in the course of the SLIM study we describe it briefly below.

Poles of attraction: *uncertainty* and *entrenchment* in technology development

Stewart (1999a) coined the term 'poles of attraction' to describe the way in which particular visions of future technological systems and services may be advanced to try and align the behaviour of other players (attract customers, enlist collaborators, deter competitors). Suppliers and other interested players may propose *particular conceptions* of future ICTs – to moot possible support: to orient and win commitments from potential suppliers of complementary products; to inform customer expectations, to ward off competitors and more generally to test out and shape ideas about technological futures. These poles of attraction become stronger as various players and technologies are aligned to around them, but can quickly fade if alternatives appear, or the idea, firm or technology falters. Recent examples include the concept of information superhighways (Kubicek, Schmidt and Beckert 1997), the Internet itself, the espousal of the 'network computer' or slim client as a solution to platform harmonisation problems in distributed business computing and as a challenge to an established pole in the shape of the Wintel PC, concepts of e-commerce and the 'New Economy', and latterly m-commerce and the mobile platform.

Social shaping studies have identified various factors and tendencies that may lead firms to seek to align their activities with other players, even to the extent of collaborating with competitors in new technology development, including escalating development costs, the difficulties of overcoming 'lock-in' effects around established solutions and achieving 'critical mass' around new solutions, which is particularly marked where there are strong network externalities, where technologies are deeply entrenched and where future technological paths are diverse and unproved (as discussed in Chapter 1). Various forms of collaboration emerge as an attempt to *share costs and risks* – and *reduce uncertainties* by foreclosing options in advance – rather than incur the costs and risks of fighting out standards wars in the marketplace (Jakobs and Williams 1999). These dilemmas have stimulated suppliers to seek to collaborate with other players (competitors, suppliers of complementary products, intermediate and final consumers) in developing standards, building markets and aligning user expectations. Since the commercial viability of a new technology may rely on convincing a critical mass of players to invest in it, such alignments become crucial to the prospects of success of a technology (Graham, Spinardi and Williams 1996). Technology development increasingly takes the form of a *networking* activity, involving inter-organisational linkages and modes of engagement between various suppliers and 'users' or their proxies. Firms take part in an ever-widening array of public and semi-public fora (Jakobs and Williams 1999).

Paradoxically, attempts to align co-producers and enrol future users in building markets and entrenching new technologies set into train 'risk management' strategies and manoeuvres that may compound the uncertainty and indeterminacy surrounding technological development. The efforts of various players to manoeuvre to maximise their prospects, hedge their bets and foreclose technology choices can end up by imparting a higher degree of indeterminacy and apparent fluidity to development processes.

The concept of poles of attraction draws attention to the way in which options may gain support and momentum – by aligning expectations and winning commitments and investments around particular technological routes – and may eventually become entrenched and materialised in technologies and institutions. Equally, commitments and alignments can very rapidly evaporate. Proffered poles of attraction may fail to command attention and win commitments from other players. Even where some momentum has been established this may be reversed and the option may lose support and fall back into the fermenting brew of emerging technologies. In this way the future of emerging ICTs is being to some extent fought out in advance in a virtual space. Many of the key decisions affecting the future prospects of a technology may be taken in the 'virtual space' of standards-setting committees and of industry fora, where key players seek to align expectations around their particular offerings.

Poles of attraction, thus conceived, are complex entities. We can see close links between this concept and the idea discussed above of technological metaphors – not least in their operation as boundary objects (see Chapter 6). Indeed, it is their multiple overlapping meanings that make poles of attraction effective as a tool for orienting thinking across diverse constituencies. To succeed a proposal must offer something for various different groups. For example, when proposing a new technology it is necessary to articulate ideas not just about proposed technological functions, but also about how technologies will be used, who will use them and who will own them.

These kinds of learning by regulating, through the articulation, promulgation and adaptation of metaphors proved important in understanding the evolution of digital experiments. As we see in Chapter 5, these metaphors served as resources and templates shaping design – in terms of specific ideas of best practice, for example for digital cities and public information services – as well as more diffuse ideas and shared understandings (genres) about how new media might be used and useful. And, as we see in Chapter 7, they helped shape the configuration of ICT projects and the relationship between the project and its context.

Having defined the concept of social learning and discussed its intellectual development, we can now turn to explore its relevance to understanding our detailed research findings in relation to the design, implementation and use of

ICT. In the next chapter (Chapter 4) we examine the processes and space for social learning. We then turn in Chapters 5 and 6 to pull together our findings regarding the substance of social learning.

NOTES

1 Cultural studies' analyses of appropriation tend to focus rather singularly upon the symbolic aspects (see, for example, du Gay et al. 1997), while work from technology studies has been criticised for retaining a slightly instrumentalist bias upon the concrete use of the artefact (MacKay and Gillespie 1992).

2 Thus the concept of domestication can be seen as being carried out not just by individuals or households, but also by institutions and other collectives, even nations.

3 Our research project addressed the formation of the policy-setting and broader technology regime in so far as these provided the context in which the cases arose. The main empirical focus was upon the design and appropriation of ICT artefacts, rather than documenting these broader processes of learning by regulating which take place through the interaction of industry players and a wider technology culture at the global level. Put schematically, design–appropriation implies a micro-level focus and an interest in micro–meso interactions, whereas learning by regulating and regime formation are predominantly features of meso–macro interaction.

4. Mapping the Process and Space for Social Learning

THE SPACE FOR SOCIAL LEARNING: STRATEGIC OPTIONS FOR DESIGN – APPROPRIATION

In explicating the different sites and processes of social learning we have begun to define something about the broader space for 'social learning'. Our empirical studies have shown that the processes of innovation (in its broadest sense to include the full cycle from design to implementation and use) vary significantly from case to case in terms of the context, the range of players involved in different types of activity over time, and in terms of the scope for learning and creativity by particular groups of players. We can represent this graphically. It is in this sense that we use the term 'space', while at the same time using it as a metaphor for the scope for diverse actors to exercise choice; to behave reflexively (Bessant 1983). We can then map the strategic options for innovation as 'paths' through this space.

In this way we can distinguish a number of different (empirical or ideal-type) models of innovation.

The linear model – a misleading ideal-type

We can conveniently start by examining an those prevalent innovation models that we wish to criticise – and in particular, the ideotypical linear, technology-driven model of innovation. Figure 4.1 provides a graphic representation of such a linear model, which sees technological *development* as limited to a constituency of design/RTD specialists and leading to stable artefacts, which is followed by separate phase of *diffusion* of these artefacts through the market to satisfy user demand. The criticisms of this model are well rehearsed (as discussed in Chapter 1). The obvious criticisms are that these stages are not separate; that technology supply does not yield finished artefacts already configured to existing need, but rather that there is an iteration between supply and use and that uptake involves an active appropriation – a learning process – rather than the passive model conveyed by the term 'diffusion'. A range of actors is thus involved. This is why we insist that Figure 4.1 represents an ideal-type – a model that is flawed and unrealistic – rather than an empirical instance of innovation.

Another weakness of the linear model is that it is frequently associated with a mechanistic, view of innovation outcomes, informed for example by supplier rhetorics and claims of the predictable and planned impacts a new technology will deliver. We can see this, for example, in the many case studies of successful technology implementation offering 'snapshot' before and after accounts of the immediate expected consequences of the supply of new technology artefacts. Instead, a *dynamic model* is needed which sees technological innovation as an interactive and iterative process involving repeated cycles from the design and development of artefacts and their implementation, appropriation and use. In short, the linear model overlooks the uncertainties and range of interactions that surround both the evolution of artefacts and their social/organisational outcomes. We will try to capture this in later representations.

Figure 4.1 is, however, useful here as it lays out something of the socio-technical space in which social learning in the generation and appropriation of ICT artefacts may take place. It thus provides dimensions against which the diversity of strategies and options for technology design–appropriation can be displayed. We will briefly discuss some of these strategic options in turn. We first examine the concept of *user-centred design*, which seeks to build a more comprehensive understanding of users and their requirements into initial systems design. We then lay out the *appropriation model*, identifying both innofusion and domestication in the implementation and use of designed artefacts. Our empirical studies of digital experiments illustrate both of these processes and throw up two further types: an *evolutionary model*, in which design evolves in the light of feedback from trial usage, and a *'pick and mix'* model in which the end-user knits together particular technology configurations from an array of off-the-shelf components and tools. Though the user has virtually no direct engagement with the design of these standard components, the latter model can be seen to afford the user considerable choice in relation to their local configurations. This thus represents a distinctive, user-led model of innovation, albeit one that is diametrically opposite in presumptions and approach to the user-centred design model.

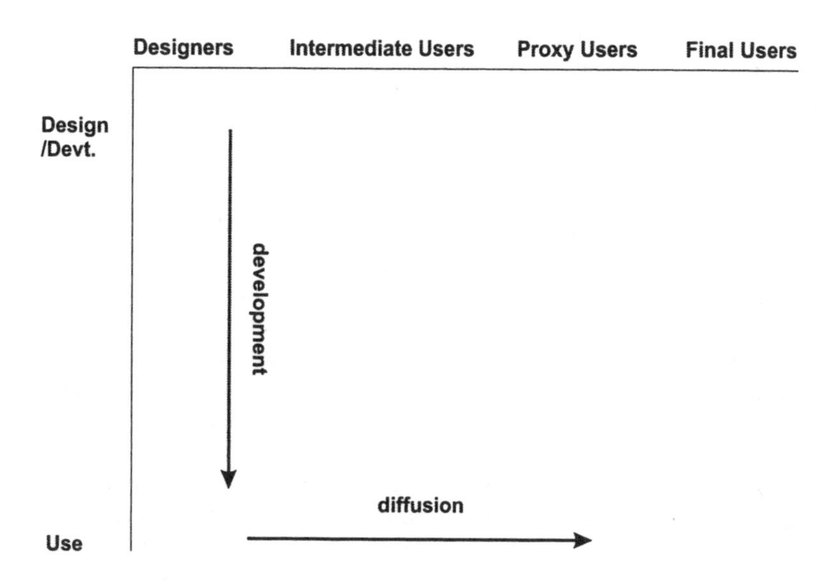

Figure 4.1 The Linear Model of Innovation

The user centred model – its origins; critique of the design fallacy

One of the main initial responses to the 'technology design problem' – the perceived failure of ICT offerings to match the culture and requirements of 'final-users', particularly in relation to workplace applications – has been the pursuit of user-centred design. The failings commonly encountered with newly developed ICT systems were attributed to the shortcomings of dominant 'technocratic' design approaches: the difficulties experienced by computer scientists and engineers in capturing user requirements, their narrow functionalist understandings of the tasks being automated and their lack of understanding the culture and practices of users. Initiatives were launched to develop richer understandings of the context and purposes of the user and to build them into technology design. New design methodologies and models were proposed. Often user-centred design involved the deployment of social scientists alongside technology developers to study user contexts (e.g. through the application of detailed ethnographic research methods [Anderson 1997]) or to bring user representatives into the design process directly. Some interesting work has been done (see, for example, Ehn 1988, Bødker and Greenbaum 1992, Green, Owen and Pain 1993). This model is represented in Figure 4.2.

However, this kind of project seems to have had only a modest influence over system design overall, and there have been problems – for example in relation to the uptake and wider applicability of models that emerged and, more profoundly, in their failure to generate distinctly different models of artefact. A number of the digital experiments we explored have tended to resemble this kind of project – particularly those with broader ambitions to be exemplars and to fulfil certain social objectives rather than narrow pragmatic goals.

Whilst the shift towards user-centred design represents a significant and positive development, we need to avoid the pitfalls of what we have termed the *design fallacy*: the presumption that the primary solution to meeting user needs is to build ever more extensive knowledge about the specific context and purposes of various users into technology design. In large degree the shortcomings of this view arise because the emphasis on the complexity, diversity and thus specificity of 'user requirements and contexts' (and the consequent importance of local knowledge about the user) is taken up within an essentially linear, design-centred model of innovation to emphasise the need for artefacts to be designed around the largely unique culture and practices of particular users. By seeing computer artefacts as largely fixed in their properties, and thus privileging prior design (Procter and Williams 1999), the key question becomes one of building ever more extensive amounts of knowledge about the context culture and purposes of users into the designed system.

However, recognition of the complexity and diversity of user settings does not necessarily imply that technological design should or will be entirely shaped around the detailed needs of particular users.

We return to this matter in the next chapter. In relation to the discussion in this chapter we note, in particular, that the configuration and customisation of cheap, generic component technologies has proved a remarkably effective way of acquiring ICTs – and one that has, arguably, had far more impact on ICT systems as developed and used than user-centred design. What is critical in terms of this current discussion is that the development of configurational technology is a model in which *technology design and implementation are closely coupled*, with a strong resemblance to the evolutionary development models outlined (see figures 4.4 and 4.5).

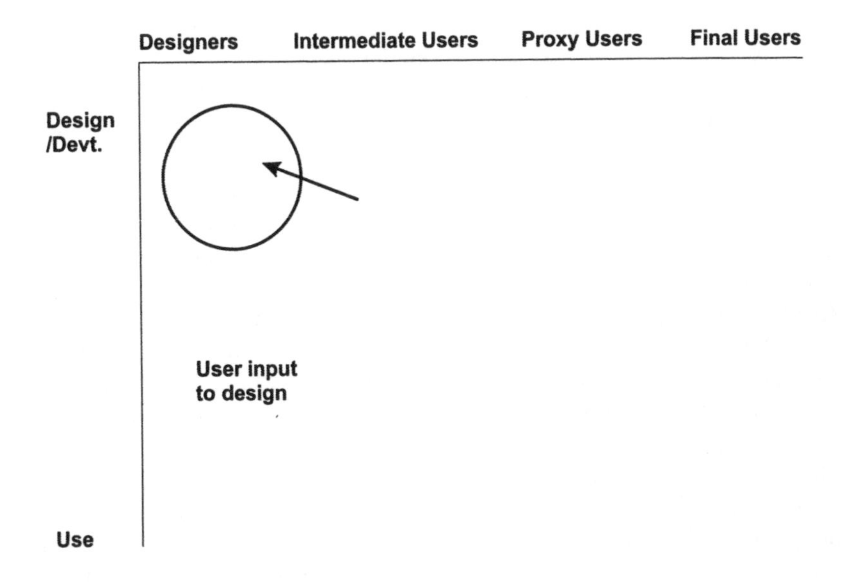

Figure 4.2 User-Centred Design

The appropriation model: innofusion–domestication

In the innofusion model, the boundary between technology development and implementation/use becomes eroded. The implementation arena is an important site of innovation. Technology applications emerge through an iterative process involving various kinds of feedback from the implementation and appropriation of artefacts to technology design and development. The innofusion concept (Fleck 1988) draws our attention both to the struggle to get things to work in a particular context, which may transform both the artefact and our understandings of its utility and use, and the possibility of feedback to future technological supply. This is shown in Figure 4.3.

Through this process, 'users', in their efforts to apply and use technology offerings, become actors in technology design – directly contributing to the reconfiguration of their own systems and potentially providing resources for future technology supply. The latter case, of course, depends upon the existence of channels to relay implementation experiences to suppliers and users of technology elsewhere. Fleck's innofusion model was articulated in relation to industrial automation, where supplier and user are large organisations with substantial financial and technical resources, and where there is a contract and direct contacts between engineers and managers of both organisations. Even in these circumstances, there was often a failure to collect and utilise implementation experiences (Fleck 1988, Webster and Williams 1993). The problems of supplier–user links are obviously far greater in relation to mass-produced consumer goods distributed through the impersonal market.

In Figure 4.3 we show innofusion and domestication as separate cycles. Their location indicates the different centre of gravity of these processes, respectively, around the technical elaboration of an artefact, particularly in the implementation phase, and around the elaboration of practices and meanings in consumption and use of an artefact.[1] However, we should not see these as being different and necessarily separate kinds of activity. Both of these kinds of social learning activity take place to a large degree simultaneously in the course of the appropriation of an artefact.[2] The distinction then is not primarily in terms of the kinds of activity – but in terms of its centre of gravity and in the kinds of outcome that are being analysed. The distinction is, however, worth making in that it does highlight differences in social learning potential between different settings. It reminds us that different groups of actors may be drawing different things from an innovation process (e.g. suppliers may be enhancing their offerings; domestic consumers, individually and collectively may be learning whether and how an artefact may be relevant to their purposes).

Our case studies provided ample evidence of innofusion. These are discussed in the next chapter. Many of the digital experiments involved users in trialling new ICT offerings. Digital experiments often go beyond user-centred design by involving token or proxy users in more or less natural settings – giving them space to experiment and develop practices and usages around trial applications. This was particularly notable in relation to the applications of ICT in education (van Lieshout et al. 2001) and within organisations (Buser and Rossel 2001). Innofusion was less in evidence, however, in relation to the 'cultural content' products for more mass markets (Preston 1999), where there was a bigger gulf between the development and appropriating constituencies. In this case other intermediaries played a greater role in understanding in bridging between these worlds (i.e. in terms of developing an understanding of the potential market of users and how it might be exploited). We discuss the role of these 'appropriation intermediaries' in more detail below.

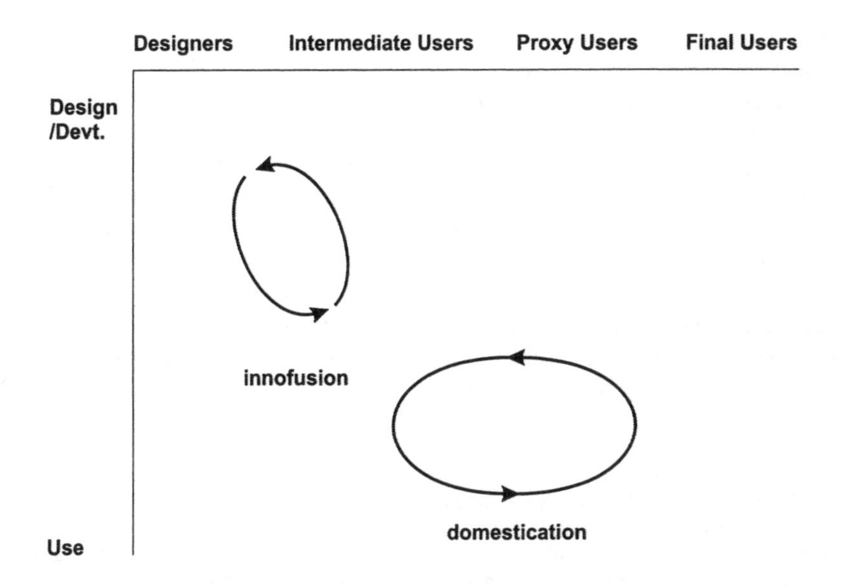

Figure 4.3 Innofusion and the Domestication of Artefacts[3]

The evolutionary model

We can use this way of mapping the space for social learning to explore the various ways in which digital experiments were conducted – and the different locales for innovation and social learning in terms of which actors were involved at which stages and around which aspects of the configuration of a system.

One important model emerging from several of our case studies, but perhaps best exemplified by the Amsterdam Digital City (DDS), is the *evolutionary model*, in which there is a permanent process of experimentation (encompassing both innofusion and domestication) as the project was developed and progressively became entrenched (see Case Summary 4.1). Rather than design being conducted and completed prior to and separately from an implementation and appropriation effort, both activities were conducted in parallel on a more or less continuous basis. An important part of the evolution of DDS and the development of content for it was the way that the technology enthusiasts sub-culture which first developed the concept were willing to open their project up to other constituencies: initially groups involved in media and cultural products and, furthermore, especially latterly, including non-specialised users as information producers as well as consumers. The constituency of players involved in development thus broadened in the course of building the market for its products. The project never finally stabilised; technological development never stopped. Instead the project has undergone a process of permanent evolution in which innofusion and domestication are combined (van Lieshout 2001).

Figure 4.4 shows both the involvement of wider ranges of players in design and implementation in digital experiments, and the extension of this strategy in the 'evolutionary development model', in which developing the technology, and building the market go hand in hand. In the evolutionary model these are continuing activities and there is no clear boundary between technology development and diffusion. In the case of DDS we saw a progressive broadening of the socio-technical constituency of involved players; barriers between technology developers and users were eroded, and the boundaries of the project expanded.

Case Summary 4.1
Amsterdam Digital City DDS – an evolutionary model?

DDS was somewhat exceptional among the digital city initiatives we examined as it was the earliest such development, and has established a substantial and self-sustaining information system. An interesting feature of the DDS case is the way in which it has managed to simultaneously involve both expert and non-experts – further technical innovation of the system has proceeded hand in hand with the appropriation of DDS by non-experts (both as users of the system and producers of information pages). For example, the interface for DDS went through three distinct technical stages, (see The Interface Design for Amsterdam Digital City – Case Summary 5.2)

What was perhaps more important about DDS (which distinguishes it from many other digital city initiatives which have not achieved the same momentum and market size) was the way in which project development was opened up to a range of groups. DDS has been successful in enrolling a critical mass and sustaining this 'market' of information producers and users thereby ensuring the viability and survival of the experiment. Why has DDS become entrenched more effectively than the hundreds of digital cities that followed it? Our study was not able to identify a single cause for this success; instead what seemed to be important concerned the *process of development* – and in particular the openness openness of the experiment in its conduct; the heterogeneity of enrolment of diverse actors that was achieved by a 'cascade of activities that together created the network in which the digital city became grounded' (van Lieshout cited in Lobet-Maris and van Bastelaer 1999). In this process domestication and innofusion take place *in tandem*. A project and concept of a community information system that started off amongst a sub-culture of politically motivated technology enthusiasts has been opened up to range of different constituencies and cultures (notably to actors and groups from media/cultural product circles, and to non-specialists in the community [van Lieshout cited in Lobet-Maris and van Bastelaer, 1999, van Lieshout 2001]).

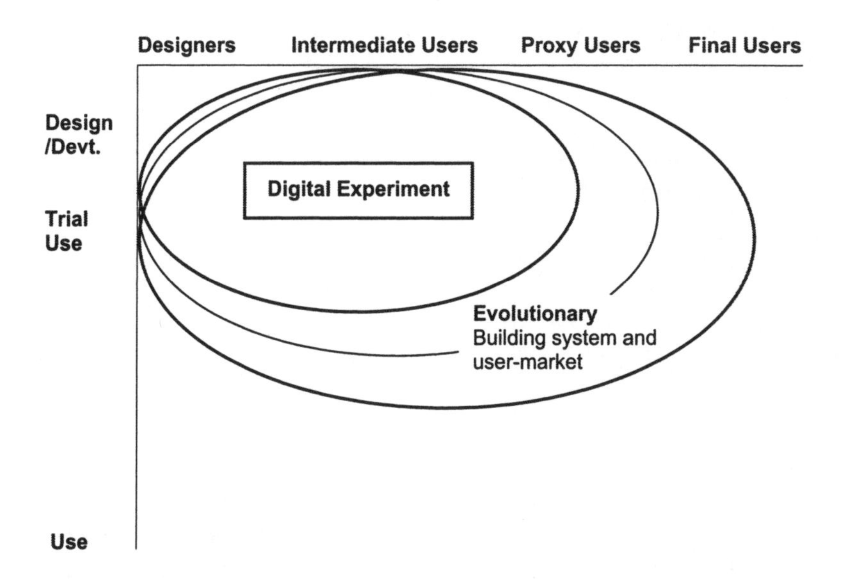

Figure 4.4 The Evolutionary Model

The 'pick and mix' or laissez-faire model

Figure 4.5 shows the 'pick and mix' laissez-faire model of technology appropriation. ICT applications are increasingly taking the form of 'configurational technologies' (see Chapter 1 for a fuller account of this concept) – deploying an array of standard and customised element on a 'pick and mix' basis to meet particular local purposes and exigencies. This has been particularly marked with the emergence of the World Wide Web and Internet-capable PC as the basic platform.

The majority of the digital experiments identified in our project involved the development of applications from largely standard technological components – typically today the Internet-capable personal computer and the World Wide Web (but also, for example, the CD-ROM). Although some of our longer-running projects (DDS and the Scottish Community Information Service) started off with more limited, dedicated, specialised text-based communication facilities (for example based upon traditional bulletin board technologies), they were mainly forced to migrate to the Web in the light of its growing popularity – both in relation to the wider array of services and sites thereby available, and the growing expectations of users for its attractive and readily understood user interface. The success of the Internet/World Wide Web model was rooted not only in its ease of use and open standards but also in the ease of production of websites and web-based services which made it a feasible proposition for low value content and for a vast range of information providers.

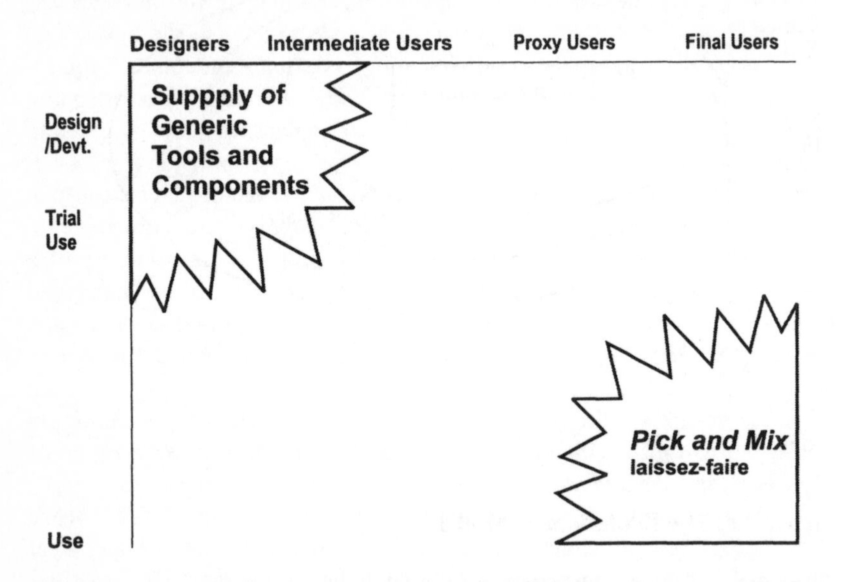

Figure 4.5 The 'Pick and Mix', Laissez-faire Model

The power and flexibility of this platform, and attempts to make it more transparent for non-specialist users, also open up the possibility for final-users who have relatively modest technical skills and resources to configure systems for themselves. We saw this across a range of other cases, from the experimental uptake of desk-top-video-conferencing in UK Bank (Procter, Williams and Cashin 1999), to the uses of the Web and email that teachers in Cable School devised for themselves.

This can be seen as a rather interesting kind of *user-led model* of innovation. The users involved in these specific cases have virtually no direct involvement in the innovation of the generic component technologies – their links to technology supply are almost entirely indirect and mediated through an impersonal market (there may, of course, have been prior forms of user engagement with the suppliers involved in the development of these generic component technologies and tools, but, even where such engagement occurred it would almost certainly be with wholly different sets of users) – however, these users seem to be able to exert considerable control over the final configuration and deployment of their systems. Lobet-Maris (1999) therefore describes this as a 'laissez-faire' strategy in so far as it is market mediated, but user led. This is, of course, a radically different interpretation of the concept of user led than we would find in a classic user-centred design initiative or 'social experiment', in which non-expert users are directly involved as actors in system development (Jaeger and Qvortrup 1991, Jaeger, Slack and Williams 2000). However, when we recognise the active and creative nature of innofusion and technology consumption/domestication – and particularly when we address these processes over multiple overlapping product cycles, rather than limit our view to a snapshot view of a single product-cycle – then we see this pick and mix strategy, shown in Figure 4.5, as one of a wide range of ways in which a broader range of players (beyond technical specialists and corporate decision-makers) can contribute to technological innovation. The prerequisite for this is the availability of a range of component technologies and tools that can readily be configured together to meet user requirements. In relation to the Internet, for example, we see the emergence of a range of developer tools created for system designers and a broader community of user-developers offering different levels of complexity and control and which are becoming increasingly accessible to those with modest technical skills. These offer a range of choices within standardised protocols and information structures.

It is important not to overstate what can be achieved by these technologies, nor to underestimate the skills required, nor the technical constraints, problems and difficulties likely to be encountered. However, our study found evidence for a novel innovation model here with some distinctive characteristics. This model further points to wider changes in the character of ICT applications and their development activities, captured by the concept of configurational technology (see Chapter 1).

Assessing these different social learning models

Overall we find that social learning around innofusion and domestication of ICT applications takes place in many different settings – and this brief overview has pointed to the very different ways in which this can be ordered and organised. The motive of this mapping exercise has been to illustrate the diversity and richness of social learning rather than to suggest that one way of carrying out such learning is 'better' than another. However, the act of making such a taxonomy also invites the idea of assessing the differential effectiveness of these models in different settings. In what way, and on what basis, could we assess the strengths and weaknesses of these different models?

Traditionally, the success of projects has been assessed in terms of whether a project proceeds to roll-out and becomes viable. From such a perspective, perhaps the key question in assessing these development strategies concerns *how well these attempts to organise social learning within particular experimental settings prefigure, match and are sustained by the wider, 'organic' social learning processes that will take place when a new product or service is launched.*

However, the social learning perspective seeks a broader approach to consideration of innovation strategy. Thus, for example, a key question, from the point of view of the viability of a new system, concerns whether users can be enrolled quickly enough and in sufficient numbers. This is not to reduce the concern with social learning to questions of market acceptance: as we argue in Chapter 9, the **rejection** of a product could provide valuable lessons. Our mapping exercise is motivated precisely by a concern to demonstrate the variety of ways in which lessons may be drawn. For example, the gulf we have noted between technology design and use could be addressed in a number of ways; one interpretation might focus on exploring how social learning processes could help identify ways in which supplier offerings could be more closely configured to the expressed needs of existing (e.g. experimental) users. However, it could equally suggest the need for the design of a product to be more generic to make it appeal to a wider range of potential users while seeking to give them more opportunity to intervene around its final configuration and use (i.e. design for configurability/further social learning). The social learning perspective does not favour particular approaches – on the contrary it draws attention to alternative strategies. The test, then, is whether social learning in experimental settings will be reinforced by wider *de facto* social learning following launch. Can this organised social learning make the 'learning curve' for new technologies more effective? Of course, what actors and analysts mean by effectiveness may depend upon their role and commitments. The goals of some may be to shorten and intensify learning; for others the priority may be to broaden the range of concerns or include new actors. (This dilemma underpins a dichotomy – that we explore in Chapter 7 regarding the conduct and management of digital experiments – between what we call the mode of experimentation and the mode of control.) And different actors may be pursuing different goals and knowledge outcomes. We return to this point in Chapter 9 where we explore the policy implications of this study and present some conclusions regarding transferability and the exploitation of digital experiments. But now we turn to examine the organisation of social learning in terms of the interactions, negotiations and flows of knowledge and experience between heterogeneous arrays of players.

THE ORGANISATION OF SOCIAL LEARNING

The key role of intermediaries and constellations

As our studies showed, the development and use of ICT products and services always involves an interchange between players separated to a greater or lesser degree in time and space and, most significantly, in terms of their institutional location, expertise and commitments. One important separation, which we have already examined, may clearly be that between the developers and users of ICT artefacts. However, we may also find important differences between technical developers (e.g. in computing and telecommunications) concerned with creating the delivery system and the implementers and operators of these systems, who are often producers of complementary products (their 'intermediate users/consumers' – i.e. the providers of content and services) that will run on this infrastructure. Thus in order to supplement their services we found the Irish broadcaster RTE collaborating with Real Audio, and Compuflex buying into other technology companies. The intermediate users will often include what ICT specialists might describe as 'domain specialists'. In our various case studies these included, teachers, TV and printed-media specialists, museum curators and public administrators. These groups also have their own institutions, cultures and traditions.

Social learning thus starts with the ways in which players, with complementary capabilities and domains of experience and expertise, work together in resolving the issues surrounding the development and appropriation of new technology-based products and services. This practical exercise in working together and solving problems (learning by doing) is a crucial component of social learning. The experience of being involved in such consortia, and learning about other players, may be as important a benefit motivating actors to take part in an ICT project as the formal goals and expected outcomes of the project (Collinson and Molina 1998, Curry 2000, Nicoll 2000).

Perhaps the central issue for this study thus concerns how to organise this kind of collaboration. We are drawn most immediately to examine the mechanisms for acquiring, generating, enhancing and exchanging knowledge. This study has in particular highlighted the role of the formal and informal networks linking together the heterogeneous constellations of players involved in technology development and use for such knowledge flows. However, for such localised learning to contribute to wider social learning, what has been learnt must be communicated and applied more widely (i.e. learning by interacting).

This brings us on to a related question regarding how best to communicate and apply more broadly the experiences of digital experiments and trials. Many of the substantive outputs from a trial may be highly specific to particular players and settings, and hard to generalise. For example, particular technical adaptations may become obsolete with the next upgrading of the delivery system. In contrast, we observe that the most important generalisable lessons may be about the process of building technologies and institutions – again raising questions about how to apply particular experiences more broadly. However, the lessons may be hard to formalise in a way that allows their broader communication. We return to these issues in Chapter 9.

The importance of intermediaries

Collaboration and linkages are thus key to social learning. The question we turn to now concerns how to organise these kinds of collaboration and exchange. We note that ICT applications typically emerge through the activities an often *dispersed network* of diverse players (which we describe as *socio-technical constellations*) who may be more or less directly linked together. We further point to the crucial role of *intermediaries* in maintaining these networks, both in linking these disparate players together and by applying elsewhere lessons learnt in particular projects and aligning and ordering views more generally.[4] Let us explore these crucial observations in more detail.

The importance of intermediaries arises from the fact already noted that technology applications do not arise unproblematically from the emergence of supplier offerings already articulated towards the expressed and nascent requirements of current and future 'users'. Concepts such as innofusion draw attention, as already noted, to the difficulties suppliers have in anticipating adequately user requirements. These observations are particularly significant in relation to emerging ICT products and services.

Let us consider first the development of new ICT applications. These often, indeed typically, emerge at the interface between organisations with very different commitments (goals, expertise, networks). This is because the viability of a product (e.g. a new delivery system) may depend upon the availability of programmes (content) to run upon it, as well as the ability of different technology offerings to work together (Collinson 1993, Nicoll 2000). Pressures for collaboration are rooted in the incompleteness of information available to individual players, as well as economic considerations (cost sharing, uncertainty reduction). ICT applications require diverse resources and different kinds of knowledge: about the technology infrastructure, about graphics and presentation of information, about the user context. Moreover, as our study of cultural products concluded, in many of these cases, 'success' also depends critically upon various kinds of knowledge and know-hows related to content authoring, design, editing and marketing functions rather than the purely technical skills/competencies (Preston 1999).

The resources needed for developing a successful ICT application are thus virtually never all available within a single organisation. As well as the diverse knowledge required in relation to the different technical fields involved in building a particular product, some understanding is required of the complementary products it must work with regarding not just its technical features but also its users and markets. In consequence, as our research has shown, digital experiments typically cut across organisational boundaries. This was particularly evident in relation to the development of new digital media which involved the creation of both 'cultural content' and ICT infrastructure (and even with other kinds of application, e.g. web pages, ICT-based services, much of what was produced and learnt related to programming and other activities that were rather similar to cultural content). Joint problem-solving in product development often went hand in hand with more or less formal attempts involving suppliers of complementary products and even competitors, and to align and stabilise expectations about emerging products markets. We thus find that different members of a coalition may well have differing goals, commitments and motives for participation; some may be interested in a generic transferable product (e.g. a technology component or platform) while others may focus upon the specific user and application at hand. *ICT developments typically involved more or less formalised coalitions of players rather than single organisations.* There must be flows of knowledge and ideas between diverse players involved in supply and use.

This raises questions about the operation of the learning economy, regarding the strength and character of linkages, in particular between supplier and users. As already noted direct supplier–user linkages may be rather weak – particularly in relation to mass-market products. Though in certain circumstances this interaction may be *mediated primarily through the impersonal market* – through the rejection or uptake of the artefact by the user – the impersonal 'pure market' is a rather poor communication device, particularly in relation to radical new technological innovations where there is no established user-base with which to engage. One example of a mass-market product from the research was *Girls ROM* – a CD-ROM given away with a Norwegian girls' magazine (see Case Summary 5.4). In this case we saw how the developers configured the product around an existing technology platform and an established market niche to increase the probable acceptance of the product. Moreover, as we see later, though this development was driven by a particular small digital media firm, it involved collaboration with other firms in publishing and advertising.

In most cases, however, we have found a more active intermediation process – with certain players (intermediaries) acting to bridge the various domains of development and use.

The use of the term 'intermediary' should not obscure the fact that intermediation is frequently shared amongst several players – though some may play a more central role than others. Intermediating roles emerge in the course of an ICT experiment rather than following formal and predetermined structures and roles. Moreover, there may be different kinds of intermediation taking place at different levels; various types of intermediary may be nested within each other, and roles may change during the life cycle of a project, as we see below in the case of Telepoly (Case Summary 4.2, Buser and Rossel 2001). Initially, technical perspectives prevailed around the installation of broadband linkages between two remote Swiss Higher Education Institutes; once installed, initiative flowed down to the university lecturers, hitherto neglected, who experimented with the use of these facilities in teleteaching. A very similar set of observations can be made in relation to the UK Cable School project (see Case Summary 4.3 below): while Cable Co. was the initiator and key intermediary in project development, within the schools involved, individual teachers became key intermediaries in the local appropriation of these technologies.

Case Summary 4.2
Telepoly – matching learning technologies and teaching strategies

Telepoly is a broadband real-time teleteaching system, set up by the two Swiss Federal Institutes of Technology, in Zurich and Lausanne, together with the Swiss Centre for Scientific Computing, to distribute higher education curricula components in a multiplexed way. The case shows how originally educational aspects are neglected in favour of realising a technologically stable infrastructure. Telepoly was not really welcomed by many of the university lecturers, who feared a loss of autonomy and perhaps displacement, and who were not included (or at least not sufficiently) in the implementation. Only after the ATM-network was established did educational perspectives get a chance after the installation of the infrastructure created a space for experimentation by teaching staff.

The case study also shows the importance of a proper adaptation of teaching strategies to the requirements of a distance teaching environment. Different strategies are possible – successful examples include one lecturer who placed teaching materials on the Web, and another who adapted a more demonstrative teaching style suitable for broadcast. Only teachers who are willing and able to adapt their teaching to the constraints and affordances of the new medium appear to be successful in their teaching. (Buser and Rossel 2001).

Particular individuals often emerged as intermediaries in the course of ICT projects – though this role was not always formally recognised, and often was not a result of a formal designation of roles, but instead arose in the interstices of organisational structures. The unplanned character of much social learning was particularly apparent in the education studies where, across a number of cases, it was clear that teachers were crucial as intermediary users and intermediary producers of ICT applications, and could be the driving force of multimedia innofusion particularly where they had space to manoeuvre (van Lieshout, Egyedi and Bijker 2001). However, their importance was not recognised at the outset, as various cases (Telepoly, Bornholm and the Cable School) all show.

ICT intermediaries can be seen, in some ways, as another incarnation of roles such as 'project champion' and the 'project manager' roles which have come to be seen as central to the successful management of technological change. However, our observations go beyond this concept. What seemed to be critical to their success in intermediation is their ability to mobilise knowledge and resources within and outside their organisation/department. In our studies of the organisational implementation of ICT applications we found that this often informal role was particularly crucial in the early 'pre-project' stages, where no formal commitment of resources had been made. For example, in UK Bank, the Research Division had to enlist sponsors from the Bank's business departments to gain legitimacy and resources to set up a video-conferencing trial by pulling together potential users and external players at a time when there was no budget and no authority to proceed (Procter, Williams and Cashin 1999). What is crucial to this intermediation role then is this ability to cross different spaces – between different organisations and different departments within organisations – and between different knowledge communities (especially that between the technical specialists and non-technical communities of providers from service and cultural sectors and users). That these kinds of informal relationships – that cut across boundaries within and between organisations – are often less durable and less well resourced than communications within established lines of organisational hierarchy has important consequences when we come to consider how social learning may be motivated and supported.

Intermediaries and constellations in application development.

In analysing the arrays of players and their involvement in social learning with ICT, we chose the metaphor of *socio-technical constellations* as it conveys the idea not only of close proximity and linkages between sets of actors, but also that there may in addition be gulfs between players with some individuals and groups remaining rather peripheral, distant and loosely connected. This concept seemed more appropriate than the more established terminologies of socio-technical constituencies (Molina 1989), systems (Hughes 1983) and actor-networks (Latour 1986, Callon 1991), which perhaps unintentionally convey a sense of closely linked and coherently organised arrays of players with direct relationships with the other players and with common commitments to particular facts and artefacts. The concept of socio-technical constellations thus points to the heterogeneity and loose coupling that we encountered in many of the contemporary digital experiments and ICT initiatives.

Further, our empirical studies reveal two particular types of socio-technical constellation:

- in *application development* – in the learning economy – involving suppliers and **intermediate** users
- in *appropriation* – concerned with technology consumption and use.

The great majority of the cases we studied were of the latter sort; centred around ICT appropriation and involving only a modest technical development effort, and often based upon largely standard technological offerings.

The relative infrequency of projects with a significant technical design/development component is only to be expected in so far as developments of new delivery systems and platforms would tend to be large-scale projects and therefore relatively infrequent, compared to the array of projects – small-, medium- and large-scale – concerned with the development and implementation of specific ICT applications.

The limited extent of technical development in those projects which involved design and development effort is a reflection of the point previously made about the character of most contemporary ICT applications as configurational technologies. When we examine the array of ICT initiatives studied in this project, we find that most of these are dealing with largely standardised technology components and systems. In a period in which the Internet and the World Wide Web became established as the dominant vehicle for ICT applications, system development, in most of these cases, was primarily a process of selecting and combining established technologies and configuring them to particular local needs. Some customisation is needed in building such configurational technologies. However, in most cases there was only a modest technical development effort.

An important observation regarding the overall array of cases studied concerns the rather *marked absence of constellations directly bringing together final-users as well as supply-side players*. One example of such a constellation is the Virtual Museum of Saint Colm Cille (Case Summary 6.1), the developers of which, Nerve Centre, operated to some extent as an intermediary both 'upstream' in the supply chain (in configuring generic ICT offerings into particular cultural products) and 'downstream' in subsequently supporting the appropriation of their Virtual Museum offerings. We explore this further in the next section.

Appropriation intermediaries

A number of the digital experiments and trials revolve around the role of certain *appropriation intermediaries* as key points of interface between potential users and new ICT products and services. They may play a particular role in *configuring* ICT component technologies and systems towards particular potential user constituencies. This seems to involve mainly a domestication effort, to make a technology relevant and attractive to individual consumers. Our studies include a number of examples, particularly in relation to cultural products including Nerve Centre and the cybercafes.

The cybercafe phenomenon is of interest – despite the rhetorics of the Information society which emphasise individual ownership and private use of ICTs (e.g. in the home), cybercafes have been formed, and seem set to continue, as collective and public spaces for different groups to gain access to information society technologies (Stewart 1999). There are different kinds of cybercafes – and even 'cyberpubs' – offering different kinds of atmosphere and geared towards different markets (Preston 1999). Cybercafes seem to play a key role here in acting as a point of access for users to new supplier offerings, and cybercafe managers are revealed as important intermediaries, promoting and channelling the appropriation process. They highlight the point that even fairly well-established ICT applications, such as computer games, are still not yet 'transparent' (Norman 1998) for the general user in two senses:

- they are difficult to set up and get to work, and
- their utility/value for the user is not yet established.

This kind of appropriation intermediary may thus play a crucial role in bridging the gap between 'generic' supplier offerings and consumers' perceptions of what they want to buy/use, which may not mesh together. From this perspective, the cybercafe managers can be seen to provide a commercially viable service not only by sharing between multiple consumers the purchase costs and risks of new games whose attraction may not be proven, but also, partly, by transferring products from one context to a more accessible one, increasing their perceived relevance and value to the customer.[5] Thus, cybercafe managers can be seen to act as important gatekeepers and brokers in the appropriation of ICT in entertainment and other everyday activities. In seeking to maintain the attractiveness of their cafes and bring in customers, they creatively deploy knowledge of both supplier offerings and user expectations and culture. They review products coming on to the market, select them and install them locally. For the customer this reduces 'search costs' (as well as reducing the risks and uncertainty of purchases). In addition, by installing it and going through the learning curve of set-up and use the manager helps reduce the barriers to access for the consumer (Laegran and Stewart 2003). As appropriation intermediaries, the cybercafe managers carry out a double practical and symbolic role, supporting the process of finding/giving local meaning as well as setting local rules for use (learning by regulating).

Our research also identified the repeated phenomenon whereby the potential contribution of certain appropriation intermediaries was not adequately recognised in advance. This was exemplified, notably, by the failure to involve teachers in relation to several of the educational digital experiments (van Lieshout, Egyedi and Bijker 2001). This neglect of social learning opportunities and outcomes and of the key players involved can be seen as a consequence of the rhetorics of technology supply – the idea that supplier offerings are already available as finished solutions, able to fulfil user requirements and to 'do the job'. This overstatement of technological capabilities undermines the need for local experimentation and adjustment. We also see in this process a privileging of some kinds of expertise and experience (notably ICT technical expertise) over others (notably the knowledge and experience possessed by domain specialists).

These observations have potentially important implications for industry strategy and public policy. They highlight the importance of different sites for innovation in technology implementation and use (as well as the sphere of technology development) and draw attention to the contribution of appropriation intermediaries towards innofusion goals, and the feedback of appropriation experiences to future supply. For example, in Japan, Sony utilised feedback from consumer electronic stores to gain insights into their newly launched CD-I products, by talking to the retailers who engaged directly with their customers, Sony obtained rapid feedback on their changing array of new product configurations (Collinson 1993). Conversely, computer hardware and software suppliers have been rather effective in getting extensive user feedback on a voluntary basis through alpha and beta testing of early releases. This raises some interesting research and policy questions about the relative costs and effectiveness of these various avenues for social learning and regarding whether it is possible to identify other effective, but low-cost channels+ for such feedback in relation to ICT offerings (which might merit appropriate public policy support if the circumstances did not favour voluntary participation). These are questions we return to in Chapter 9.

What motivates social learning? How it is shaped by its local context?

Having identified the importance of intermediaries and collaborative networks to the development and appropriation of ICT, a number of ancillary questions arise regarding how such social learning is influenced by the local setting. In particular, what factors promote this kind of exchange and the, often informal, mechanisms that support it?

Put simply, social learning will take place where it is in the perceived interests of players to collaborate and exchange information. Learning in this sense involves linking players across socio-technical space (i.e. from different industrial cultures, or between supplier and user communities). Intermediation and participation in fora for social learning involves players in costs and commitments (including the opportunity costs from foreclosing other concurrent social learning opportunities and activities). Constellations, and the various more or less informal links across organisational and institutional boundaries that they depend upon, survive because the parties are motivated to get involved and to continue to take part: in other words, they make the necessary investments because the individuals or organisations involved expect some kind of benefit.

The incentives may be complex and various. Foremost are the benefits of co-production and sharing of knowledge about ICT application development and appropriation. Such collaboration may bring *expectation of economic gain* for technology or knowledge providers (directly or indirectly, e.g. through reputation building) or similarly a *reduction in costs and risks* (e.g. from 'closure' of technological controversies; through alignment of views and enhanced social learning). The importance of reputational incentives was demonstrated in our studies of Cable School (Slack 2000a) and the Language Course tele-education project (Mourik 2001), and other cases such as the UK Videotron trial (Curry 2000) in which a key, albeit covert, motive for running a demonstrator project was to demonstrate not the technology but the supplier's capabilities, and improve the image for the organisation, and thus increase its prospects (in these cases in different, more or less closely related markets). *Positioning* in relation to future markets and reputation building thus may be important elements behind a decision to take part in an initiative – in order to demonstrate an organisation's membership of an advanced technology culture and to ensure that it figures in 'know-who' – knowledge about who is who and who is capable in the field. The diverse and complex set of incentives and purposes at play underpin our observation (below) that social learning and the role of intermediaries is a multi-level game.

It may, for example, be in the mutual self-interest of players in technology supply and implementation (especially intermediate users) to collaborate in building new markets (Collinson 1993, Molina 1992, Williams 2000). However, the situation of final-users may be rather different to that of suppliers and intermediate users. The latter are directly or indirectly bound up with the development of new ICT products and services – they may have an economic interest or be otherwise committed in terms of their areas of expertise. In contrast, many final-users may not be committed to a particular technological field, but may be involved because they have a pragmatic concern to benefit from using a technology. However, if this can be assured, they often lack an economic or professional commitment to the future of the technology in general. Thus they may well be indifferent as to particular means and technological paths – they are not interested in whether one technical solution prevails over another and they may have no generic commitment to ensure the viability of a technology beyond their own access to it. Thus, certain kinds of final-users may be in a distinctly different position to those more closely tied to a particular technological fields. Indeed, this may be a key factor underpinning the significance of appropriation intermediaries – as players with the commitment needed to act as some kind of representative for final-users. More generally this point reminds us that there may be certain classes of player, in certain application domains, or with particular relationships to system development, implementation or use, who do not experience sufficient incentives to motivate their involvement, but whose involvement may be critical (again particularly in relation to final-users).

We conclude that constellations do not necessarily emerge spontaneously. Key intermediaries may need to steer the process – and motivate the inclusion of whatever players and resources are needed. Here many of our studies pointed to the important role of public support in providing resources where these are lacking, for example from private sources, to motivate these forms of collaboration and intermediation. This finding has a number of policy implications that we return to in Chapter 9.

Our study emphasises the particularities and localisation of social learning processes. Social learning is shaped by its local context. Learning processes and outcomes differed between players according to their particular situation. In analysing the different patterns that emerged, our studies highlighted the importance of the specific 'translation terrain' – the immediate array of players with their historical and contingent concerns and capabilities, each trying to map out their strategy in interaction with other players and in the light of their broader social, economic and cultural setting. First, although all actors notionally had access to the same technical resources (in the sense that these technological capabilities were – at least hypothetically – universally available), they made different constructions of these opportunities in the light of their particular local setting and their commitments and concerns. Second, in contrast to the common predictions of a progressive alignment of ICT applications around certain global models, our studies emphasise the specificity of local translation terrains. This was demonstrated with particular clarity in relation to the digital city experiments (Lobet-Maris and van Bastelaar 1999). The emergence of Amsterdam Digital City (DDS) has provided a very powerful exemplar of successful development. However, the digital city projects that have emerged in its wake have been many and various. Quite different constructions have been made of the goals and methods of development of these initiatives.

Social learning is a multi-level game

Following on from these observations, another matter that became apparent in the course of in-depth investigations of the conduct of particular digital experiments was that the participants were often involved in a complex and *multi-level game.* As well as the formally stated shared goals of a projects, particular participants had certain more or less covert goals, which were not necessarily shared.[6] Furthermore, the various players involved tended to have a wide range of commitments and interests in a project at a number of different levels. For example, a technology supplier might have several goals: to forestall and warn off competitors, to align expectations of potential customers, to signal competencies and establish a reputation as a future player in this and related markets, which could be more or less closely geared to what they might expect to learn from taking part in an experiment. For example, the Cable Company promoting the UK Cable School project was seeking to advertise its competencies as a supplier of computer networks to the local education authorities in the run-up to a new information network acquisition.

Case Summary 4.3
Cable School – technical infrastructure at the expense of the social dimension

This service was developed to connect a number of British schools to the Internet through the use of cable modems. Cable Co., the supplier of the modems, worked with the schools to develop both infrastructure and content. The service had a number of specific constituencies beyond the schools being served: it was regarded by Cable Co. as a showcase of its capability to offer a cable-based infrastructure for local councils as well as a trial of cable modems for this and other services. The original concept was that teachers in different schools would collaborate in producing and sharing web-based teaching worksheets.

The development of Cable School stressed the technical at the expense of the social – providing an infrastructure but not establishing the time space for teachers to meet. This had implications for trust and commitment of teachers – who lacked the confidence and motivation to submit their worksheets to the system, especially when they had often copied them from commercial texts. However, the teachers used the infrastructure to develop the service in unanticipated ways. Teachers, finding that there was not a great deal of content supplied within the Cable School project innovated, used the Web connection to cache potentially useful web pages for later use within the school. The infrastructure was used to communicate – primarily via email within schools as opposed to between schools – in a manner unanticipated by the originators of the service. The system developers assumed that teachers would create a professional community between schools, when in fact it was principally used within the existing community. We have described the continued use of the infrastructure despite the lack of content as 'the social learning of independence'. The project has implications for current ICT policy in education in that it appears that the stress on infrastructure at the expense of social dimensions is being replicated (Slack 2000a).

The Language Course project was similarly developed largely as a marketing and public relations exercise by an information technology (IT) supplier (Mourik 2001). The Norwegian PTT, Telenor, supported and shaped FRIHUS 2000, an IT highway service in Fredrikstad, as a way of ensuring that it was considered as a player in emerging local information systems, particularly in the development of telecottage and telecommuting centres

Social learning may also be taking place at a number of levels around different issues. For example, the launch of a digital experiment or pilot may be motivated by a concern to address the technical and operational problems in getting an infrastructure to work as well as a desire to try out particular applications or popularise certain service concepts. Learning about complementary players and about how to work with them in developing and maintaining new ICT services may be equally important (Curry 2000). In other words, the key lessons may be more generic than the particular technical outcomes that are the formal goals of a project. For example, in Nicoll's (2000) study of an interactive TV experiment, media players sought to understand better the world of on-line services, and the issues involved in collaborating with organisations with very different backgrounds and concerns. Social learning is not restricted to acquiring the knowledge needed to build new artefacts, but includes learning about other players and how to interact with them around technology development.[7] It provides an opportunity to find out who the significant players are, what their goals are and what can be expected of them in a new and evolving domain; to build representations of these other relevant groups and to test and revise these representations. It includes building shared visions and representations, or recognising fundamental differences in goals and culture that may preclude productive collaboration. Social learning is about the process as well as the substantive outcomes of collaboration – which in some cases may arise through a structured formal collaboration and at other times may emerge in an ad hoc and informal manner.

Given these considerations it is therefore essential to stress the complexity of goals and interests against which particular projects may be developed.

Crucially, many of the important lessons may not have been anticipated at the outset. The social learning framework highlights both the uneven trajectory of technology developments and the unplanned, and to some degree unplannable, character of the learning outcomes. One consequence is that assessing whether a project can be deemed successful becomes rather a complex and subtle matter. As we see below, many of the lessons may be covert – and, indeed, may remain tacit to those involved – particularly in relation to the strategies of building constellations/experiments. This also flags the role of intermediaries in carrying and applying these local experiences more widely, as *brokers* of knowledge and influence. We thus see the intermediaries not as passive transporters of knowledge but as creative, reflexive actors in two important ways: first, in their ability to encapsulate particular experiences and identify which elements can be applied more widely and how second, in their ability to identify other settings in which learning experiences can be applied (which may be rather different from the context in which the experiences were acquired). An important issue may thus arise on the conclusion of a project. If a project is wound-up, some players may be forced to move to other organisations or areas of work: sometimes they may become involved in a related or similar project; at the other end of the spectrum of outcomes they may find themselves in a situation where there is little prospect of applying their experience. A crucial question in relation to the broader and longer-term social learning outcomes thus becomes whether the players involved will continue to have an institutional context that provides the resources and motive to carry these ideas forwards subsequently. For example, problems may arise, particularly in relation to public-funded experiments, if the constellation and some of the key actors involved in the project exist only for the duration of the experiment, and are dispersed and move to wholly different settings when the project comes to an end. This has important policy implications, which we discuss in Chapter 9.

In assessing social learning in a digital experiment we also need to take into account different timescales and contexts. To assess the significance of a digital experiment we must address the experimentation and social learning about the design and use of ICT products that may take place within the project. We return to this point in Chapter 7, which focuses upon the conduct of digital experiments and highlights some key choices in the organisation of social learning. However, we must also examine how these may form part of a broader set of learning processes around ICT products and services over a longer timescale and within a broader setting. Though these broader social learning outcomes may often be indirect, unanticipated and not immediately apparent, they may prove to be the most important outcomes of a particular project.

The reflexive intermediary – social learning and reflexivity

This leads us to consider further the broader role of intermediaries in steering a digital experiment, and their attempts to negotiate and navigate in a complex and changing context and deal with the different exigencies that may arise. All the projects we examined evolved substantially in the course of their lifetime. For example, the key actor in Craigmillar Community Information Service (CCIS) was highly successful in obtaining continuing public support for his project, by redefining the project in relation to new funding opportunities, new technologies, new potential users of the system – for example, joining up with other information services to increase the attractiveness and user base, upgrading the system to web standards, setting up training schemes (Slack 2000b). New challenges were encountered and new activities undertaken, that were not necessarily recognised at the outset. Likewise Amsterdam Digital City (DDS) has been through three distinct launches, in the face of changes in technology and concepts of the service and how it is organised. Similarly, after successfully developing the prototype of their *Virtual Museum of Colm Cille* as a CD-ROM and website, Derry Nerve Centre embarked upon a range of activities to support the appropriation of this product in local communities and in particular schools. These points bring us back to the central and role played by the individual and organisational intermediaries in identifying and developing a response to these new challenges.

Case Summary 4.4
Craigmillar Community Information Service (CCIS) – learning to be fundable

CCIS was developed in the early 1990s to connect community groups in a deprived area of Edinburgh. To ensure its survival, CCIS has had to adapt itself to a changing environment and changing funding opportunities. The project has evolved over time, with changes in technologies, concepts of community information services and how they might be used.

CCIS has *learned to be fundable* in an uncertain environment where the priorities for funding change regularly. The uptake of the service among the original community groups reached initial targets; however, there was a need to develop further user constituencies in order for the project to remain viable and secure further funding. It has also learned to exploit technologies in an effort to enrol new user constituencies (for example, upgrading from proprietary bulletin board to web technology, partly in response to growing interest in the latter amongst the community). Recently, it has funded itself by offering training services. However, these expansions of the service are seen by some as moving on and away from the traditional community base and role of CCIS.

The critical comments from those involved in the project and in the wider community turn on the question of how far CCIS fulfils its original remit, and how far it has become diverted by technical potential and funding exigency from its basic aims. Different user constituencies may have very different conceptions of the success and failure of the system. Those involved in the project *ab origine* suggest that CCIS has become too outward looking. The original system allowed local community groups to communicate (by providing modems to link their computers together). The development of web-based interfaces allows access to external services and the presentation of cultural content to remote as well as local audiences. The manager portrays the shift from community integration to access to the Internet as a matter of social inclusion 'Cyber-rights'. Through the establishment of groups such as 'Keyboard Kids' and 'Cyber Grannies' he seeks to enfranchise as many people as possible through connection to the Web (Slack 2000b).

In considering the changing role and strategies of intermediaries it is therefore important to analyse them as *reflexive actors*, responding creatively to novel and changing circumstances Indeed, the role of the intermediary can be seen as one of providing a *focus of reflexivity* in the social learning process. This was to some degree recognised by the intermediaries themselves, who were often concerned to theorise about the world in which they operated, how it was changing, the challenges this posed and how they might respond to the consequent difficulties and uncertainties.

This kind of reflexive activity is thus key to our concept of social learning. The social learning perspective flags the importance of knowledge acquired through practice. It suggests that all players have opportunities for reflection and insight into technological change – including everyday citizens in their daily decisions about whether and how to engage with a new technology. However, certain players may have a particularly advantageous viewpoint which offers special opportunities and incentives for reflection, and for developing a more systematic understanding – for example technology project leaders, managers and intermediaries who have a broad *span of control and action*. Intermediaries are then of special interest in terms of their what they do and the knowledge and experience they acquire.

The opportunities for social learning also varied according to whether this took place in a structured manner, for example as an explicit goal of a formal collaboration, or whether it emerged spontaneously amongst particular actors in the course of a development project. This has bearing on the extent to which locally acquired experience will be applied more generally and on the level of reflexivity achieved.

We conclude this chapter with a brief aside on the relationship between (reflexive) researchers and (reflexive) practitioners. As previously noted, the idea of social learning highlights two kinds of reflexive processes:

1. the ways in which the players themselves are changing their strategies in the light of particular experiences (which are articulated into new models and programmes of ICT innovation) and

2. the possibility that social science research can help us understand how better to organise this learning. These considerations suggest a range of social learning settings and potentialities.

One focus of social learning then is with socio-economic researchers, who have access to particular skills, methods and theories, and the time and other resources needed to develop a *span of overview* to develop and refine systematic understanding of social learning processes and outcomes. However, researchers do not have exclusive insights.

When we come to consider the policy implications of the social learning perspective we find that one of the key issues concerns how to extrapolate from particular digital experiments and contexts of learning to other settings. And this is, of course, an important challenge for social scientists as much as for practitioners.

NOTES

1 Strictly speaking, consumption of course starts prior to implementation, though continuing into the stage of normal use of an artefact – in so far as the purchaser is enlisted to acquire an artefact: there is thus some prior engagement around the projection of supplier promise and the generation of user expectations.

2 We considered (on the grounds of simplicity of presentation) the idea of defining innofusion as the technical appropriation and domestication as the cultural appropriation of an artefact. However, we explicitly reject such a formulation as it implies a separation between 'technical' and 'cultural' that is impossible to make in practice and unhelpful for analysis in that it goes against our emphasis on the *socio-technical* character of many of these processes. Though this simplistic dichotomy certainly captures something of the differing primary concerns and foci of the exponents of these concepts, both concepts were articulated with a broader view. The boundaries between innofusion and domestication are thus by no means clear-cut.

3 In these diagrams, locations on the two axes are indicative rather than exact. Thus no implication is intended in the horizontal dimension in terms of the fact or sequence of involvement of particular players such as intermediate or proxy users in every case. Nor is there a presumption that there will always be some kind of trial use between development and use. However, there is an implication that projects which do roll out to widespread use will move from the top left to the bottom right of the picture. These figures then seek to convey something of the different routes that may be taken in this process.

4 The concept of intermediary, of course, appears in industrial economics, in studies of the development of network technologies such as inter-organisational IT networks (Williams 1995, Graham, Spinardi and Williams 1996) and in cultural studies (in relation to cultural intermediaries – whose role is bound up with the circulation of ideas). The SLIM study seeks to draw on the different dimensions of intermediation thereby elucidated.

5 We can thus see appropriation intermediaries as playing an important role in reducing the costs and risks of acquisition of new products which may impede the

uptake of a new product and, if a 'critical mass' of users is not established, which may even threaten its commercial viability.

6 Although the diversity of commitments and perspectives may present problems for a consortium – see, for example, Nicoll (2000), it should not necessarily be seen as a weakness. Given the dangers of premature alignment of views about a technology (see, for example, Howells and Hine 1993) it could be argued that development constellations need to contain a wide variety of different groups, with their own goals and criteria for assessing success.
7 Schot and Rip (1996) highlight two forms for learning in collaborations: broad learning, i.e. learning about each other, and deep learning, initially focused on the task in hand, subsequently a more reflexive form of learning based on linking and clarifying values.

5. Social Learning in Technology Design

The preceding chapter explored the *process* of social learning – the mechanisms involved and how they were shaped by their context. In Chapters 5 and 6 we begin to explore the *substance* of social learning – i.e. what was learnt and how – in these various digital experiments. For ease of reference we have divided this discussion into two chapters, addressing, respectively, social learning in the design process (and how the user was represented in this process), and social learning in the appropriation of artefacts, in their implementation and use (including how the artefacts and its uses were reconceived in innofusion and domestication). However, it will be immediately apparent that the implicit linearity in this schema is artificial and not very helpful. In particular, ICT applications are not developed as a series of isolated design initiatives, but often emerge in the short term through experiments and pilot applications which couple design and early implementation and in the longer term can be seen to result from repeated cycles of innovation. Many of the issues and activities therefore cut across a simple boundary between design and use.

DESIGN AND THE PRIOR REPRESENTATION OF THE USER

Design is evidently a strategic site in technological innovation. Partly as a result of its obvious importance it has been unhelpfully and unduly afforded a privileged position within several analytical traditions in technology studies and in computer science. We therefore start this chapter with a critical analysis of this traditional treatment of design, which is called into question by our social learning framework which addresses technology design and implementation/use in tandem.

In designing an artefact, some model is needed of the anticipated user and the ways in which the artefact will be used. At the heart of much of this analysis has been a concern to understand these *representations* (Vedel 1994) of how the user and use of a technology are represented in design. The second section addresses this and shows how this work has sought to 'problematise' representation in several ways: by showing how particular designs have been based on erroneous views of the users' identities, purposes and contexts; by explaining why it is difficult to anticipate the users' needs and requirements; and by proposing methods of improving design by building a more effective and realistic representations of users. We criticise the 'design fallacy' – the presumption that the best way to improve the design and usability of ICT artefacts is by building increasing amounts of knowledge about specific users/contexts into prior design. We then articulate a social learning model of design/representation which attempts a more realistic account of design processes as operating within particular social settings. System design, we argue, is inevitably based upon incomplete knowledge, uncertainty and inconsistent views about future users and user settings, and must therefore resort to methods of generalisation and extrapolation from this and other design contexts – metaphorical leaps – that are potentially contestable and may be contested. A more nuanced understanding of design of ICT artefacts is advanced. Design is seen as a process of configuration in the sense of being not an open process but substantially constrained by its historical and institutional context. We further identify a number of dilemmas surrounding design strategy, for example, regarding the extent to which solutions should seek to cater for the diversity of user settings or should be generic. Design is also constrained by prior technology design and acquisition decisions – particularly in a context in which ICT applications take the form of configurational technologies.

POINTS OF DEPARTURE: A CRITICAL REVIEW OF APPROACHES TO DESIGN

Technology is bound up with the idea of fulfilling human purposes more cheaply, efficiently and effectively. Technologies such as ICT are often designed and developed with particular activities and objectives in mind – for example an organisational ICT application is geared towards achieving particular kinds of organisational change. And they are sold to individuals and organisations on the basis of meeting our needs. Failures of technology to deliver expected improvements are often attributed to failings in system design, in turn rooted in shortcomings in the ways in which the activities and purposes of users – that the system must support – are captured, for example, in system requirements specifications. But how do these purposes become articulated and embedded within technology artefacts? The design process is clearly an important phase of the product development cycle, in which many elements of a technology are open and amenable for influence.

There is a rich seam of analysis that we can draw upon here from technology studies and other fields – notably socio-technical approaches to computer systems design such as the fields of Human Computer Interaction [HCI], participatory design and Computer-Supported Cooperative Work [CSCW]). However, to develop our own line of argument it is important first to make a critical assessment of some trends within this literature and flag our points of departure.

The issue of design was central to the emergence of technology studies as an area of debate and field of study. Starting from an the identification of some of the undesirable social and environmental implications of newly emerging technologies, critical socio-economic analysis moved on to ask 'what was giving rise to technologies that were having these effects?'. Critiques were advanced of the dominant form of technologies developed (MacKenzie and Wajcman 1985). Their aim was to undermine the taken –for granted aspects of technological innovation, and to highlight the choices underpinning these apparently 'technical' (in the sense of neutral, rationalistic) design decisions. For example, the 'social shaping of technology' perspective sought to investigate the choices inherent in technological design and development, and to show how these were influenced by the various values and interests involved. The main current of social shaping and constructivist studies of technology has thus typically conceived design as shaped by the values and context in which technologies are developed. Implicit in many of these early accounts was a particular and rather 'politicised' view of technological design as purposive – in the sense of readily able to capture the particular priorities of powerful actors and of artefactual design as being *richly informed* by specific values, concepts of the user and use. This provokes questions about how these values are embedded, and how they can be identified.

The archetypal 'social shaping' study by Noble (1979) sought to establish this by pointing to the explicit intentions of the developers (of automated machine tools), and the suppression of one technological option (Record Playback) in favour of another (Numerical Control) – alternatives which offered clear (apparently self-evident) implications for the outcomes in user organisations. Noble's machine tool case refers to competing technological design choices in a highly polarised and visible context of choice (conflicts about the role and productive contribution of craft machinists versus engineering technicians, in a context in which the perceived industrial strength of craft labour was the focus of managerial concern). A similar conception of the significance of design underpinned the espousal by socially concerned engineers of alternative approaches to technological design (for example in ideas of Human-Centred Technologies and participatory design [see, for example, Ehn 1988]).

Whilst deeply critical of contemporary models of modernity these accounts retained modernist presumptions regarding how the social context shapes the content of design and how these values are reproduced – in which the design of the artefact is a more or less simple reflection of the values and priorities of designers and developers – values which are assumed to be reproduced (or at least favoured [Winner 1980]) when these artefacts are deployed and used.

Rather different methodologies may be needed to grasp design choices in other, less starkly delineated contexts, in which more complex arrays of actors and objectives may be at play. Important insights have emerged from analysts with roots in discourse theory. Thus Akrich (1992, 1992a) and Akrich and Latour (1992) claim that we may interpret the endeavours of designers as efforts to inscribe certain preferred programmes of action by users in the design of a given artefact or technological system. Designers visualise a script of preferred reactions to the artefact, and they try to shape the technology in order to make these reactions as mandatory as possible (Sørensen 1997). Akrich (1992: 208) defines the concept of script in the following terms 'Designers thus define actors with specific tastes, competences, motives, aspirations, political prejudices, and the rest ... A large part of the work of innovators is that of "*inscribing*" this vision of (or prediction about) in the technical content of the new object. I will call the end product of this work a "script" or a "scenario".' In the same volume, Akrich and Latour (1992: 261) propose the specific vocabulary of '*prescription; proscription; affordances allowances*: 'What a device allows or forbids ... from the actors that it anticipates; it is the morality of a setting, both negative (what it prescribes) and positive (what it permits).' Akrich is interested to address the collision between the anticipated user inscribed in the artefact and the actual user.

In a similar vein, Woolgar (1991) describes designers as seeking to 'configure the user' – in terms of the characteristics of the user and how they may respond. By 'setting parameters for the user's actions' (Woolgar 1991: 61), the behaviour of the user is configured by the designer and the user is disciplined by the technology. In this sense the technology (and the designer) constructs 'the user'.[1] This concept has been widely taken up.

Paradoxically, this kind of critical analysis has tended to produce a rather 'essentialist' (Wajcman 1991) account of how the values and objectives that surround design become embedded in the artefact – and can thus readily be 'read-off' by the analyst – and, second, to convey a somewhat mechanistic 'linear' view of how those embedded values and scripts are likely to be reproduced when those artefacts are subsequently consumed. In particular, Woolgar's 'configuring the user' concept, perhaps unintentionally, conveys a distinctly unidirectional concept of the outcomes of technological change that would be more characteristic of technological determinist accounts.[2]

We argue that many of these early accounts have yielded a simplified, 'stylised' narrative in which technological design/development has somehow favoured the interests and values of particular actors and values over others. What results is a rather over-politicised account of technological development, viewed through the perspective of a particular value system; an account that still bears the imprint of its intellectual roots in terms of the continuation of a view of design that tacitly retains some modernist presumptions regarding how the social context shapes the content of design: which social values and relationships become embedded in design and how, and the outcomes when the artefact is deployed. Across a number of studies a somewhat stereotypical 'received approach' seems to emerge in terms of the theoretical presumptions and the research design/methodology adopted. Thus many of these studies (e.g. Akrich 1992, 1992a, Woolgar 1991) have tended to presume that design incorporates a comprehensive representation of the intended users, their purposes and the context of use (for example in relation to the users' skills, their identities, e.g. in terms of gender [Cockburn and Furst-Dilic 1994] or other social features such as race and class) and the activities that may be seen as appropriate. Researchers starting from this perspective have then tended to look for the problems that may arise where that representation is restrictive or out of line with the actual users that emerge or can be anticipated to arise (for example highlighting the problems where engineers have relied on their personal experiences and presumptions to articulate a rather unrepresentative model of the user [Berg 1994b, Nicoll 2000]). In other words the design problem arises because the user has been misconstrued; the wrong voices and the wrong values have been catered for, and other users and perspectives overlooked. Underpinning this is the tacit presumption that the design of the artefact is a more or less simple reflection of the values and priorities expressed in the design setting and in particular the priorities of designers and developers. In this respect, designers are often treated as if they were acting with a high degree of autonomy – as evinced by Latour's (1988) modern princes or 'Sartrean engineers' – thereby ignoring the extent to which design operates in a broader organisational setting and is conditioned by broader structures, relationships (e.g. the organisation, inter-organisational constellations) and tools through which the perspectives of users and other actors are articulated and represented (Mackay et al. 2000).[3] The attempt to identify and analyse critically the way in which users are represented may be motivated by a desire to give voice to those who may have been marginalised and excluded from development choices. This is, of course, one of the central tenets for technology studies. The problem with this kind of approach is that the desire for narrative clarity is at the expense of recognising that design takes place in a clamour of voices (external, internal, internalised) and is inevitably an imperfect compromise amongst a multiplicity of factors and actors.

Of course, this tendency stands in contrast to another important current in technology studies which has emphasised the flexibility of interpretation of technology (Pinch and Bijker 1984). Of course, the above writers from a discourse theoretic background (e.g. Latour, Akrich, Woolgar), see technology as a 'text' that is capable of different readings (even though their writings convey a strong sense that the technology inscribes a preferred reading). The work of Akrich (1992, 1992a, 1995) in particular draws attention to the differences and interactions between 'the designers projected user and the real user' (1992: 209). She advances the concept of de-scription to highlight the way (illustrated in her case study of a technology transfer project) that an object may be reinvented and reshaped in use.

We point to a growing body of recent accounts, influenced by developments in cultural and consumption studies, which portray consumption as an active and creative process (Sørensen 1994b). These emphasise that, although the designer may seek to prefigure the user – and thus implicitly to constrain the ways in which the product is used – ultimately users still retain flexibility regarding the meanings they attribute to technologies, and over choices about how the artefact will be appropriated. For example, in relation to household technologies this involves choices in terms of where the product is located and how it is incorporated within family routines (Morley and Silverstone 1990). This domestication of technology often involves innovation by the consumer – using technology in ways not anticipated by the designer (Berg 1994a). The social learning perspective draws upon and seeks to integrate both of these analytical traditions within a view that locates design within broader processes and multiple cycles of technology design and appropriation

THE 'SYSTEM DESIGN PROBLEM' AND THE 'DESIGN FALLACY'

Another important line of critique of design practice has centred around the perceived failure of ICT offerings to match the culture and requirements of users and in particular of the 'final-users' who must operate the system. In parallel with the preceding debates in technology studies there has been a continued focus upon the difficulties in capturing 'user requirements' adequately in computer system design, which were seen to underpin project failures and the repeated experience of a gulf between expectations and outcomes achieved when new computer-based systems were introduced into commercial and public organisations. Traditional requirements-capture techniques applied by computer scientists and engineers, which emerged from the successful automation of routinised high-volume record-processing tasks in early commercial computing, only yielded narrow, functionalist understandings of the tasks being automated. They failed to produce a sufficiently detailed understanding of the intricate purposes and activities of the growing array of 'end-users' as computers were introduced across the organisation and geared to supporting an increasingly wide range of often complex and variable organisational activities (Friedman 1989). The result was a gulf between designed systems and the circumstances and practices of the various groups of potential and actual users.

Requirements capture is a potentially difficult problem because the needs of various current and potential users, are not fixed entities, already fully articulated in the user's mind, that can simply be 'read off' by systems analysts and designers – e.g. simply by asking the user what they want. Many users find it very difficult to specify their requirements and are often poorly placed to anticipate how new technological capabilities might be deployed. This is one of the reasons why needs, and the means by which they may be fulfilled, evolve, partly in the face of new technical capabilities and practices. In the face of the recognition of these problems, a range of user-centred design initiatives was launched which sought to develop richer understandings of the context and purposes of the user and build them into technology design. New design methodologies and models were proposed, including structured system design methodologies (from those particularly concerned to manage development more tightly) and (from those particularly concerned to emancipate the development process) participatory design and human-centred design.

User-centred design often involved the deployment of social scientists alongside technology developers to study user contexts; there have also been efforts to bring user representatives into the design process directly. Some interesting work has been done that sought to develop exemplary models (see, for example, Ehn 1988, Bødker and Greenbaum 1992, Green, Owen and Pain 1993). However, with hindsight, this kind of project seems to have had only modest influence over system design overall, and some serious questions can be raised about their effectiveness – most immediately in relation to the uptake and wider applicability of models that emerged from user-centred design initiatives. More fundamentally, it can be noted that user-centred design initiatives largely failed to generate distinctly different models of artefact to those emerging from conventional design settings. This was true also of some of the digital experiments explored in this study that had broader ambitions to be exemplars and fulfil certain social ambitions.

Participatory approaches to system design have achieved wide currency within computer science training and practice. Ethnographic and, in particular, ethnomethodological approaches were advanced as being uniquely suited to addressing the intricacy of specific contexts and practices – identifying the crucial differences and distinctions that conventional requirements capture techniques would all too easily overlook (Anderson 1997). There are obvious limitations to the role of ethnography as a central method for requirements capture,[4] though ethnographic approaches could be used to supplement other methods or as an alternative to 'user participation' in design, in which representatives of various current and potential users could express their requirements for the new system and contribute directly to requirements specification and to design and development decisions.

Whilst the shift towards participatory and user-centred design represents a significant and positive development, we need to avoid the pitfalls of what we have termed *the design fallacy*: the presumption that the primary solution to meeting user needs is to build ever more extensive knowledge about the specific context and purposes of an increasing number and variety of users into technology design. In large degree the shortcomings of this view arise because the emphasis on the complexity, diversity and thus specificity of 'user requirements and contexts' (and the consequent importance of local knowledge about the user) is taken up within an essentially linear, design-centred model of innovation to emphasise the need for artefacts to be designed around the largely unique culture and practices of particular users.

If we were to see computer artefacts, once designed, as largely fixed in their properties, and thus privilege prior design (Procter and Williams 1999) the key question becomes one of building ever more extensive amounts of knowledge about the context, culture and purposes of users into the designed system. Following on from this, socio-economic research, and in particular ethnographic studies of users were proposed to identify the *right* values and overcome the user requirements problem, by capturing the increasing amounts of knowledge about specific groups of users and their purposes, practices and thus requirements that could be incorporated into the design of the artefact.

In this respect it parallels the weakness evident in early technology studies accounts of design, sharing with them what we could describe as a 'heroic view' of design – in the sense that designs are portrayed as finished products inscribing particular views of the user, user activities and priorities into the artefact in a way that will be fulfilled in subsequent use of the technology – which at the same time demonises engineers. The design-centred model, with its exclusive preoccupation with prior technological design ('the design fallacy'), can be criticised on a number of grounds:

- It is unrealistic and unduly simplistic.
- It may not be effective in enhancing design/use.
- It overlooks important opportunities for intervention that are revealed, for example, if a design-implementation life cycle model is adopted that, for example, addresses unanticipated use and the unpredictability of innovation outcomes (see Chapter 6).

In particular we argue against the model of design as an inductive process of accumulating ever more information about current user requirements. Recognition of the complexity and diversity of user settings does not necessarily imply that technological design will or should be entirely shaped around the detailed needs of particular sets of users.

The creeds of many computer science development methodologies and the discourses of the design fallacy emphasise the need for a closer fit between designed systems and the specific requirements of actual and intended users – and the consequent importance of finding means to articulate those requirements and building systems more tightly around them. However, we can readily identify a number of reasons why systems may not be designed around specific user needs.

1. *Design is inevitably generic*. First, as we argue below, artefactual design is inevitably generic to some degree in relation to specific users. Even where users have been directly involved in system specification, a process of selection and synthesis of requirements, of *decontextualisation* (Schumm and Kocyba 1997), must take place to abstract and generalise from specific needs.
2. *Factors may favour standardisation*. Second, there are many factors which may mandate in favour of generic or standardised solutions, rather than capturing the specific, probably idiosyncratic, requirements of particular configurations of users. There are important trade-offs in software supply and acquisition strategies between making artefacts unique versus some level of standardisation. For example, the cost and other benefits of reusing software 'code' may allow user organisations to acquire more powerful and dependable packaged solutions at a fraction of the price of bespoke solutions. We thus find a trade-off between the *increased utility* to the particular user and *higher cost* of solutions custom built around their particular requirements and the price/performance benefits of cheaper generic solutions which may match user requirements less exactly (Brady, Tierney and Williams 1992, Fincham et al. 1994,). Users may choose to adapt to the constraints of cheaper packaged software.[5] These trade-offs yield a range of technology supply strategies (as illustrated in Chapter 1 – especially Figure 1.1). Thus the rapid spread of packaged software and tools reminds us that the possible price (per unit functionality) advantages of mass-produced standard solutions may outweigh the costs to particular users of adapting systems or adapting their activities to system constraints and affordances. The attractiveness of standardised offerings is further increased by the possibility of combining them with customised elements into configurational solutions. This is further

assisted by conscious attempts to design such component technologies to be readily linked together and customised.

3. *Designed artefacts as unfinished; as resources for social learning*. Third, but linked to the preceding points, we articulate a differing view of designed artefacts not as finished solutions for user needs, but as *resources for subsequent social learning*, including innofusion and domestication and further enhancement of future designs.
4. *Other actors have requirements too*. Finally, user requirements capture is only one amongst many considerations underpinning design. There are many other factors, including development costs, the existing technical infrastructure that a new project must link in to and standards. In practice this means that designers of particular artefacts and systems are typically interacting with an array of other players with particular concerns as well as user requirements. System design is a markedly heterogeneous activity. Indeed, if we adopt a broader view of design as the arena in which purposive choices are made about the configuration of an artefact, the work of many other players could be seen as constituting different parts of the collective process of design, including, for example, marketing specialists and senior managers as well as technical specialists.

This last point highlights a difficulty with the term 'user' itself. The concept of the 'user' is a very specific representation those people with which one has a relationship though the provision of a technology or a service, one that is particularly used by technology developers. Other professional groups, such as doctors, public representatives, commercial executives or teachers will use terms that have very different symbolic value, such as patient, client, student, voter, citizen, audience or customer, which represent the particular relationship they have with those they have service relationship with. These non-technology focused players will always place their relationship and concerns above those of the technology developer, and may be concerned maintaining existing relationships with those people or organisations that are taking on a life as 'users' of a new technology. Design may often fail when designers do not get to grips with these alternate representations and priorities.

We will return to these, and a number of related arguments, in the remainder of this chapter.

DEVELOPING A REPRESENTATION OF THE USER

In designing and developing an artefact some model is needed of the anticipated user and the ways in which the artefact will be used. These *representations* (Vedel 1994) of the user/use may be more or less specific. Designers do not simply develop an artefact with particular functionality – they must at the same time develop some conceptions – both tacit and explicit – about the kinds of uses that may be attempted, about who the users may be and about the contexts of use (van Lieshout et al. 2001).

Why then does it seem to be so difficult to develop robust and effective models of the user at the outset?

Considerations of efficiency and emancipation would appear to favour a turn to the user – that more knowledge should be acquired about actual and potential users their contexts and purposes to inform representations and design. There are however a number of reasons why it may be difficult to acquire such knowledge.

1. *Extent of supplier–user linkages.* The manner in which developers are able to engage with users this may depend upon the kind of linkages that exist between technology developers and intended/potential users. These opportunities for linkage between developers and users vary substantially between different kinds of ICT project, depending for example upon whether it is a novel application or an extension to existing products/services; whether it is geared to a specific user group (e.g. members of an organisation) or to the mass market. For example, design in the application of ICT within an organisation involves a tightly structured context, with direct linkages between supplier/designer and user, and the possibility of engaging end-users directly or indirectly in system development. At the other end of the spectrum, direct linkages between designer and user may be absent or attenuated in the case of mass-market products.
2. *Radical/disruptive versus incremental innovation.* In this context there may be particular problems with radical innovations – radical in the sense that they are not incremental enhancements of a product or service already in use. In these circumstances the user does not yet exist. To put it another way, we are uncertain about the applicability to a new setting and new product of the knowledge we possess about current users and uses of existing products.
3. *Knowledge of existing users may become outmoded.* These uncertainties in assessing user requirements are amplified because the needs for a particular new product or service are not fixed and predetermined but emerge through a dialogue between currently expressed user

requirements and emerging technical opportunities in the course of the domestication of a technology. We know that needs will evolve.[6] We therefore know that existing user representations, even if strongly supported by empirical evidence of actual user behaviour and priorities, are likely to be called into question.

4. *Uncertainty about identity of future users.* This links in to another uncertainty about the *identity* of the future user – which will only emerge through this engagement with future artefacts. Issues thus arise about how to target the future market – about how to define and put boundaries around particular classes of user. Providers may make presumptions about their key markets – and the priority attached to particular market segments – which may be called into question. This was exemplified in our study of 'The Den on the Net' (Kerr cited in Preston 1999), the set of websites developed by RTE, the Irish national TV and radio broadcaster, to 'add value' to its traditional media offerings. The subsequent experience after these had been launched, emphasised the importance of the 'diaspora' market comprising Irish nationals abroad and emigrants (in particular in the USA and Commonwealth nations) which provided a potentially enormous market of people already using the Web and interested in 'Irish culture' (see Case Summary 6.7). The existence and importance of this particular market could in principle have been anticipated, though it was overlooked in the case. However, it is by no means always the case that the potential market/future user base can be adequately prefigured at the outset.
5. *Uncertainty about identity of virtual users*. Indeed, when considering on-line applications the very idea of cyberspace emphasises the profound uncertainties about the identity of the user. This was, for example, one of the key dilemmas facing the digital city initiatives and community information systems. On the one hand, there were 'grounded users' – those located within a particular physical community (or other interest group) whose characteristics were more or less well known, but which were potentially knowable. However, there were also visitors from outside the physical community, whose identities remained uncertain. Issues arose about how or whether to cater for them. Should they have the same status and access as grounded users? Different approaches have been adopted. DDS in Amsterdam, for example, distinguishes between 'residents' and the presumed casual access of 'tourists'.[7]
6. *Difficulties in assessing users/uses in private spaces.* Some settings for ICT use may be less readily observed and known by suppliers. Designers and developers have had particular problems in addressing

the needs of potential customers of domestic technologies as use occurs within the private sphere of the home. The responses of domestic ICT users (and refusers) are shaped in the complex social dynamics of the family (Morley and Silverstone 1990, Silverstone 1991) with its particular culture, values and 'moral economy'. They are not an amorphous and homogeneous mass; their responses are differentiated by gender, generation and class. However, there is relatively little systematic data about ICT usage in these settings. Given the paucity of links between designers and potential users, designers have often relied primarily upon their own experience and expertise; starting from their understanding of technological opportunities and imagining how these might be taken up in their own households – which may be far from typical (Cawson, Haddon and Miles 1995). This may cause problems in the acceptability of ICT offerings. One obvious example today is the 'baroque' design of most contemporary video cassette recorders – which the vast majority of consumers find very difficult to use. Similarly, suppliers lack understanding of 'the housewife' as a possible user, and of 'her' needs, and have had little appeal to many customers (Berg and Aune 1994) For example, 'smart house' technology demonstrators reflect technology push rather than user need; they have not really addressed the realities of domestic labour (Berg 1994b).[8] An important factor concerns the level of competence available to the ICT application developer. Levels of in-house expertise in areas with tools and concepts needed to analyse user requirements and settings, such as marketing and product design, were extremely variable between suppliers. Smaller providers were often lacking in this respect. In contrast some of the larger and more established players had been building up their base of knowledge and expertise, and, for example, possessed rather extensive knowledge about their current customers (for example media companies such as RTE in Ireland know a great deal about their established market and were able to apply this in designing their web-based offerings).

7. *Users are heterogeneous.* The final point is that it is unhelpful to talk about 'the user' since users are extremely heterogeneous. Even where there is direct information about the actual users of a new product there may be uncertainty about how reliable it is as a basis for extrapolation. For example, we already know that early adopters may be technology enthusiasts with very different expectations and requirements than later adopters (Rogers 1983, Norman 1988). Technologies designed around the presumptions of technology enthusiasts in design and early adopters may exclude the rest of the market. Issues thus arise about how to segment the intended market – and thus about how to cater for various classes of user (and which classes may implicitly or explicitly be

> overlooked) – or about how to evolve a user representation/product design as the user market evolves. These considerations all serve to reinforce our central thesis about the profound difficulties involved in developing an adequate understanding of the user: the ambiguities about the identity of the users (rooted in difficulties in anticipating intended and actual future users of a system) and the difficulties in acquiring and validating information about their characteristics and requirements (difficulties in knowing 'the user' stemming both from the aforementioned difficulties in anticipating future users and from uncertainties about the validity of existing knowledge about current users/users and its applicability to future users). Despite these difficulties in building representations around direct evidence of actual users, representations of 'the user' did emerge. Indeed, representations of 'the user' may perhaps be, at least in part, a corollary of the fact that the actual users are elusive and ambiguous and hard to know. 'The user' is, thus, a metaphorical construct, that may stand in for a number of actual and imputed players and roles. This observation in turn forces us to ask other questions: Who is being represented? Who is the representation for? For example, it would appear that 'the user' is invoked in the first instance from a designer/developer perspective.

We can analyse the representation of the user as a boundary object (Star and Griesemer 1989 – see also the discussion of metaphors and boundary objects in Chapter 6) in the sense that they allow communication across diverse groups with differing perspectives and may also serve to order these differing understandings. For example, for an industrial ICT application the concept of 'the user' could refer to a number of people involved and implicated in different ways in an eventual system: the user organisation, its senior managers involved in a decision to buy, the specialist and managerial staff utilising the information it contains, the 'end users' operating the system (and perhaps even the receivers of final services). Which of these potentially competing representations of the user are articulated, and for which audiences or purposes. Particular constructs may reflect a political need for certain images of the user (as evinced by Nicoll's [2000] concept of 'users as currency') – whether reflecting particular exigencies within the developer setting or the priorities of the sponsor or other external players. Elision between constructs or the substitution of one by another may have important consequences – for example where various end-user perspectives are subsumed within a generalised view of managerial requirements and use of a system. Indeed, the intense interest in user representations is precisely inspired by a concern that certain groups and their requirements may be overlooked or submerged, resulting in systems that exclude or fail adequately to address the requirements of particular groups of potential users.

These issues about user representation came up very strongly in our case studies. Later in this chapter we analyse the ways in which particular development contexts shaped both the internal and broader politics of the design setting and its consequences for user representation in a number of our cases.

Design as an hypothesis about the user

Our study sought to avoid making specific presumptions at the outset about the way in which design-representation is conducted and sought instead to pursue a detailed empirical understanding of these matters. We took the broad view that artefacts embody something of a *hypothesis about the user* (Lobet-Maris and van Bastelaar 1999). In this sense, digital experiments and trials can be seen as providing an opportunity to *test* these hypotheses.

In many of our case studies in which a novel ICT product/service was being developed there was no existing user base and very weak or no prior linkages between designers and users (for example in the establishment of novel community information systems). Across a range of these case studies, we found that in the design and development process for ICT products and services, these hypotheses about the user and use often remain implicit and under-specified. The presumptions made about the user were typically largely unstated and often poorly elaborated. This finding presents something of a challenge to the 'heroic' technology studies tradition on design and 'the design fallacy' that we criticised above.

Representation

It remains the case that development/design needs to prefigure a number of elements about the context, purposes and activities of the user – to develop some kind of *representation.* Representation relates to a number of different elements. For example, Nicoll's (2000) *contextual usability* model conceives the usability of technology as a complex of interdependent elements within a particular context, including *usefulness*, (the development of) *usage* patterns, and the particular social and cognitive exigencies of situated *use*. Developing this further we suggest that representation encompasses: the technical configuration of the system, content, usage, uses, rules (formal and informal) about proper usage/users.

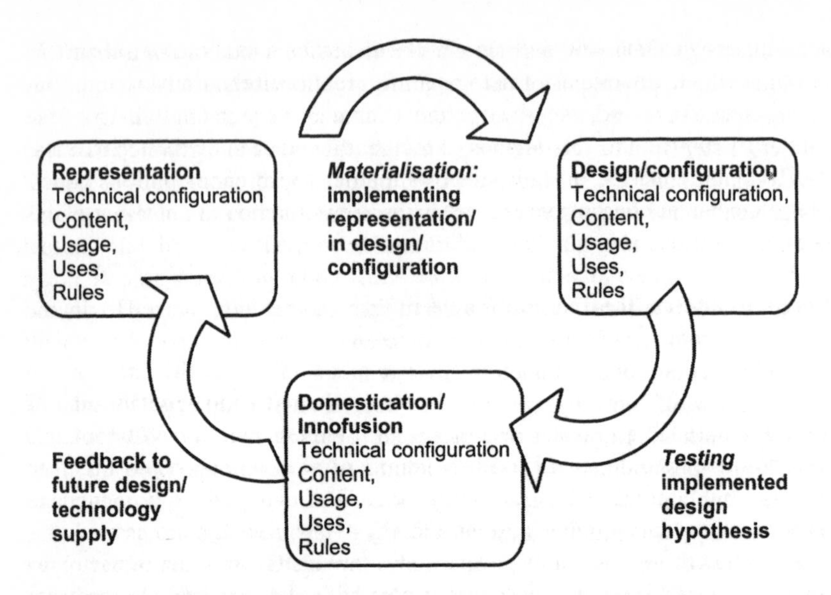

Figure 5.1 Schematic Diagram of User Representation and Appropriation

Figure 5.1 incorporates these ideas into a schematic diagram, laying out our model of the cyclic relationship between user representation, design and appropriation, based on our integrated social learning model. It shows that design is not a one-off act, but is part of an iterative series of activities, informed by earlier design practice and feedback from the appropriation and use of other systems (earlier technologies in this application domain; similar technologies in related domains). Figure 5.1 shows schematically the relationship between representations in design/development and appropriation – highlighting the various elements involved, and the iteration between the materialisation of user representations in particular designs, the testing of these design hypotheses in the domestication/innofusion process and feedback to future technology supply/design.

In principle, the hypotheses embodied in the design/representation are attempts to prefigure these very dimensions of the eventual use of the artefact. However, it is often difficult, indeed impossible, to prefigure these reliably – hence the importance of social learning, both in testing the design hypotheses and for feedback from appropriation to future technology design/representation (Akrich 1995).

There are then, a number of elements through which designers may seek to 'configure the user' (in the Woolgarian sense – comprising both attempts to prefigure/incorporate the user in the design of the artefacts and to align actual users to that view). Vedel (1994) identifies a range of mechanisms through which developers seek to configure the user – advertisements, directions for use and technical guides – as well as technical design. Van Lieshout (1999) takes this further to include the context of the technology. Applying these ideas, in their study of the design of digital cities, Lobet-Maris and van Bastelaer (1999) identify the different elements of the artefacts through which the user may be configured:

1. The user interface: for example in the choice of design metaphor and the way information is presented in the interface.
2. The language and terminology used in the interface.
3. Services offered – and the types of information available.
4. Rules allowing or forbidding particular behaviours.
5. Access possibilities – e.g. where equipment is accessible, opening hours etc. imply different types of user.
6. Training courses and instructions.

These can be illustrated in our study of the Copenhagen Base information system (see Case Summary 5.1) which revealed a number of specific ways in which users were configured.

Case Summary 5.1
User Representation and Configuration in Copenhagen Base (CB)
The analysis of CB revealed a number of specific ways in which users were configured around more or less explicit representations, including:

- *Language*: the decision only to use Danish in CB served to exclude many potential users such as foreign visitors. The information is structured around the domains of responsibility of Copenhagen Council rather than for example classes of users or their priorities.
- *Types of service and information content*: fact that CB offered city information but not the additional attractions of free email and other interactive services offered by other digital cities.
- *The rules in the digital city*: CB has rules that do not permit groups with politically extreme views to speak. However, these cannot guarantee the exclusion of such users and their views. And if the chat-rooms and discussions are filled up with extreme political views or pornographic suggestions many 'nice people' will avoid the service.
- *Opportunities for access*: in Copenhagen computers in public places like libraries give access to the service. Without this, users who do not have a computer in their home or at work would be excluded from the digital city.

An assessment conducted of how the CB could be made more attractive and accessible the targeting of publicity (and the kinds of target group underpinning advertising); the type and level of skills required to use the system; and the design of training courses, guides and instructions will all influence which groups will be attracted to take part. For example, complicated technical guides speak to a different user group (privileging those with technical skills and education) than user-friendly instructions directly presented on the screen. Other groups will presumably not be attracted to the same degree.

As these studies showed, designers and developers may draw upon various intellectual resources in building a representation of the user; not just particular knowledges of current or intended users, but also more general set ideas – for example standard principles of ergonomic design; ideas about good cognitive design of interfaces, and knowledge of existing applications and methods of using them that may have gained wider currency and become a widely accepted genre of use.

Resources for design: building a hybrid representation of the user

Following on from this we ask what intellectual resources do designers and developers have for building a representation of the user? Figure 5.2 shows some of the sources of ideas and information that designers/developers may deploy.

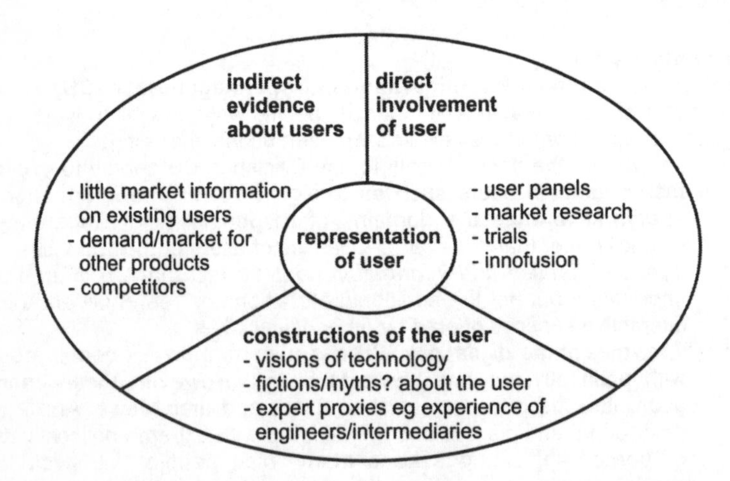

Figure 5.2 Resources for Building Representations of the User

Figure 5.2 illustrates a number of points. First, in a context in which information about potential users is typically incomplete or of uncertain reliability, players may be obliged to 'knit together' different kinds of knowledge from diverse sources and with different degrees of gearing to 'actual' users.

Second, there may be relatively little empirically grounded information about existing users. Many studies have drawn attention to the crude and limited ways in which technology suppliers have sought to understand the requirements of potential users (Cawson, Haddon and Miles 1995). Though there have been important improvements over the last decade in the techniques by which firms gain direct information about users (for example through market surveys, consumer testing of prototypes by panels of 'proxy' users, feedback from 'real' users of early versions of the product [Akrich 1995]), progress has been uneven, and the cost of collecting such user information can be very high. Many developments will take place in a context of incomplete and perhaps rather limited information. In the absence of direct knowledge of users, there may be resort to more or less 'expert' constructions of the user. Constructions of the user created by 'experts' (e.g. engineers or intermediaries) may be derived from their own personal experience and culture (Nicoll 2000) or may be more firmly rooted in experiences in this or other product markets and matched against technical capabilities. Oudshoorn and Pinch (2003) have described this strategy, following Akrich (1995) as the '*I-methodology*'.

The I-methodology has obvious pitfalls. It is very easy for designers to overlook apparently obvious features of the users and their contexts. This was amply illustrated by the design of the Language Course video-conferencing system (Mourik 2001). Although the designers knew the users would be children, they forgot one of the most obvious changes needed; the height of the computer screen. The camera mounted on the computer showed only the top of the head of the children (whose height had to be raised with pillows and books on the chair). As we see later, the design also overlooked the potential for playful use of the system by these children, as well as the possibility that the children or other members of the family might reprogramme the computer for other uses.

We should not, however, vilify the I-methodology. As well as being widely prevalent, and therefore needing to be taken seriously, it has a number of strengths. Technical specialists, through their intimacy with technical opportunities and ability to think how these could be exploited, may be effective in visualising new ICT applications (a point we return to in Chapter 7 in discussing design as creativity). The views of other experts may be an important design resource. Developers do not work in a vacuum – but may be influenced here by the behaviour of peers and competitors – which may be reflected in clustering of supplier offerings or the mutual reinforcement of supplier visions and presumptions. These kinds of alignment of expert views have potential benefits in narrowing search processes (thereby making them cheaper and quicker) and reducing perceived uncertainties, as well as helping establish generic understandings about how ICT products and services may be used (see the following discussion of metaphors and genres). Such alignments may also, of course, underpin problems of mission-blindness, which have been identified as the root of a number of high-profile and expensive failures of ICT systems. Expert constructions may be informed not only by rigorous evidence and pertinent experience but also by visions of technology and fictions (myths?) including anecdotes or stereotypes about the user which may turn out to be more or less close to actual users.

The potential weaknesses of the latter forms of 'evidence' are clear (informed by 'implicit' rather than explicit techniques [Akrich 1995]). However, we would like to take this point further. Our third argument is that all the forms of information about future users carry their own uncertainties and difficulties. For example, the most systematic empirical information available about user choices and preferences (as revealed, for example, in aggregate form through market behaviour) may exist in relation to established products. However, in a context where products are changing rapidly this knowledge may become outmoded. Expert views are liable to be rooted in prior experiences in other related markets. The question arises as to how far one can extrapolate from such information. The problem perceived in relation to 'radical innovations' is that knowledge about the user and use of existing applications is seen as not providing a reliable guide to the novel application.[9] Possessing particular types of knowledge about users, though a potentially useful resource, does not eliminate the difficulties in deriving effective representations for design.

Finally, we must remember that design representations are not simply constructed around evidence about expected/imputed users and uses of a technology, but may be limited and indeed compromised by time and money constraints. And, as we saw vividly illustrated by the cases of Craigmillar Community Information System (Slack 2000b, Case Summary 4.4) and Language Course (Mourik 2001, Case Summary 7.2) the key constraints were not necessarily derived from considerations of the user or use but were rooted in the context – the translation terrain constituted by the priorities of the collaborators and the relationships between them by the actors.

Problems in building representations on the basis of knowledge about the user

Empirically grounded information about users of a new product may be sought through direct involvement of users in panels, through market research surveys and trials. However, various difficulties arise regarding the interpretation of such direct information about users.

Even in initiatives geared towards direct 'user involvement' in the design/use of technologies there are concerns about representativeness, and about the validity of the information obtained. Thus, even where there is an attempt to sample potential actual users (e.g. from amongst current users of earlier systems) there are issues about whether the users selected will be representative of the range of ultimate users that may emerge. More fundamentally, will they continue to be representative? For example, user panels need to be introduced to new technologies and given some training in their use – their selection and training, however, mean that they are in some ways no longer independent and representative of the wider publics they are supposed to represent. Similar problems surround the establishment of new spaces, such as user panels and pilot schemes, to allow people to interact with and assess emerging technologies. Sørensen (1999: section 2.3) describes this as 'simulated social learning' involving as it does 'people that are supposed to act as if they were users, but under artificial, laboratory-like circumstances.' But how should appropriate 'proxy users' be selected? Is their behaviour in the laboratory a good basis for understanding behaviour in everyday life? Would these responses provide a secure foundation for anticipating the larger cohort of real users that the developer wants to cater for? When organisational applications of ICT are being developed it is commonplace to involve 'super-users' who have demonstrated their ability to think creatively about the enhancement of systems and working practices, and who are therefore better able to articulate their requirements. However, such users, because of their commitment and ability, may have rather different concerns and understandings and may thus fail to pick up the requirements of more typical end-users of systems (Hepsø 1997). These factors, of course, tend to mandate in favour of live trials – which would allow experimentation around the acceptance and utility of a product by a wide range of users in naturalistic settings (Nicoll 2000). These observations underpin the potential importance of social learning in the innofusion and domestication of ICT – in that it can provide rich sources of more direct and reliable information about 'actual' user responses to supplier offerings. It in turn raises questions, which we return to below, about how such appropriation experiences can be fed back to generate more robust user representations for future design.

We have identified one of the key problems in building design representations that are representative, in terms of the difficulties in specifying likely groups of emerging and future users and characterising their various requirements and purposes in ways that would adequately represent eventual arrays of users.[10] The other key problem, linked to this, arises from the heterogeneity of users – and the difficulties in catering for the diversity of users groups, their contexts and purposes. The dangers in building design around the requirements of particular, and perhaps idiosyncratic and untypical, users mandates in favour of encompassing an increasing user number and range of users. However, this will inevitably result in a proliferation of often mutually incompatible preferences and requirements between individuals and groups – necessitating some kind of selection and prioritisation. In our studies we found that, as designers tried to cater for ever wider arrays of groups of users, concrete conceptions of the user derived from particular individual users/groups tended to become more diluted, and the representation of the user became more abstract. A number of analysts have highlighted this shift in requirements analysis from various partial representations of concrete users to a generalised representation of 'the user' (Woolgar 1991, Akrich 1995, Oudshoorn, Rommes and Stienstra 2004) Instead of differentiating between many various sorts of users – each of which might have their own peculiarities – designers trying to interpret user characteristics seek to incorporate user requirements wholesale into their design in an attempt to serve them all at once. In consequence, instead of working around a particular representation of 'the' user, no user at all is conjured up. For example, the users were neglected in the design of the user interface for the first two generations of the Amsterdam Digital City (DDS). In this context, technology-driven visions tended to serve as a proxy for representations derived from knowledge of actual user constituencies; user and use seemed often to be extrapolated from technical potential (Lobet-Maris and van Bastelaar 1999). Only one group was explicitly catered for – power users with high skills – and there was little attention to the fact that this reproduced a user community typical of the Internet-using community of the day: young, highly educated people and predominantly male (van den Besselaar, Melis and Beckers, 2000, Rommes, van Oost and Oudshoorn 2001, Rommes 2002).

Case Summary 5.2
The Interface Design for Amsterdam Digital City (DDS)

The first, text-based version of Amsterdam Digital City, DDS1.0, entailed a menu-structured interface based on Freeport software. Facilities were created that enabled users to browse through the information pages, and that enabled them to communicate with each other, but these (the menus, the underlying software) were not fine-tuned to peculiarities of different groups of users. The second interface, DDS2.0, was, in words of the designers themselves, the result of a technically driven approach, as they sought to be amongst the first organisations experimenting with graphical interfaces. Only with the start of the design of DDS3.0 did users became an explicit part of the design process, in terms of a largely shared presumption that the interface should be accessible and understandable to all users, irrespective of age, gender or education. During the design process, one group of users was explicitly taken into account. Interestingly enough, this was not the group of inexperienced beginners, but the opposite group: the so-called 'power-users' with experience with a UNIX-based environment, who were offered entrance to the shell structure of the interface (van Lieshout cited in Lobet-Maris and van Bastelaer 1999, van Lieshout 2001).

Akrich (1995: 183) calls for methodologies to bring a number of user representations into design – particularly in relation to end-users – and to ensure that 'certain user representations – which would otherwise not be considered by the innovators and the entrepreneurs – are taken into account'. This is an important challenge, made more difficult by the complexity of user representations

Users are not unitary, and neither are design and development constellations. Different aspects of the representation of the same users are important for different players in the development process. Figure 5.1 suggests a number of dimensions in representation, design and innofusion: the technical configuration, the local rules, content, usage and uses, which are salient for different actors. They can also be interpreted in different ways. For example, while commercial managers may be concerned with the activities of an organisation, interface designers are concerned with activities of individuals. Representations of users are always partial, fragmented and hybrid, and open to alternate interpretation.

As we see below, there are tensions in design between differentiating artefacts and standardising them, and between aligning with existing users and their representations and transforming them. We return to this last point in the final part of this section.

DESIGN AS CONFIGURATION: EXISTING ARTEFACTS AND ROUTINES AS A DESIGN RESOURCE[11]

As we noted in our discussion of the design fallacy, above, many treatments of design unhelpfully adopt a snapshot view that privileges prior design and pays little attention to innovation in and artefacts implementation and use (Procter and Williams 1999). Linked to this, a frequent weakness surrounding many 'micro-sociological' accounts of technology design has been their tendency to treat design as an isolated episode, undertaken from scratch, and divorced from its broader history in terms of the broader development trajectories of artefacts (and their uses) across multiple design-implementation cycles. Their search for scripts, inscribed by technology designers, underplays the cumulative and taken for granted aspects of most design activity. Drawing on other analytical traditions (especially research into the design and implementation of industrial technologies and work organisation) we argue that design is rarely a process of invention *ab initio*, but is typically a process of *configuration* – in two senses:

- Design has been frequently shown to involve the application of relatively restricted sets of rules for reconfiguring – selecting, reworking and recombining – existing knowledge and practices, often bound up with the application of existing repertoires of criteria and recipes (Whipp 1985, Clark and Staunton 1989);
- It involves the creative selection and configuration together of a variety of already existing components, as well as novel elements.[12]

Workplace ICT applications have been shown to involve combinations of condensed social relations (e.g. existing work and informational practices) existing component technologies coupled together and reworked around visions of realisable technical and organisational change geared towards concepts of best practice (Fleck, Webster and Williams 1990). Furthermore, as we saw in Chapter 1 in our discussion of ICT as configurational technology, systems design and development relies increasingly upon off-the-shelf components. The development of already existing components may have been separated from current ICT application designers in both time and social space. Whilst there will necessarily be some judgement by application designers that their selection is appropriate, these imported artefacts will bring with them their own embedded functionalities and embedded presumptions. It cannot be presumed that designers (or for that matter socio-economic researchers) are in a position to read off these 'imported scripts': the origins of many of these features are likely to remain obscure. However, the emerging artefact is not just a cumulative accretion of fossilised social relations (as we might imagine some kind of geological sedimentation of its past history); innofusion and configuration processes around these artefacts often involves the unpicking, selective adoption and reworking of these already developed component technologies (Fleck 1988, Fleck, Webster and Williams 1990).

This draws our attention to the processes whereby earlier innovations became incorporated within subsequent design:

- Through the incorporation of previous design-implementation experiences from different settings. For example, specific bespoke solutions were adapted and sold on as more or less standardised software packages, perhaps incorporating an increasing array of specific options (e.g. regarding user context and purposes) as embedded options within supplier offerings (Brady, Tierney and Williams 1992, Webster and Williams 1993, Pollock, Procter and Williams 2003);

- Through the tendency for artefacts to become generic – transformed in this commodification process – to become relatively independent of the requirements of particular users or classes of user. Indeed, many ICT applications increasingly take the form of 'media', capable of supporting a wide range of activities.

These are particularly features of programmable technologies, i.e. their operation can be changed by changing software parameters (and software is cheap to reuse), and which are designed to allow some level of reconfigurability and flexibility in implementation. Moreover, a growing array of interoperability standards assists the linking together of separately developed systems.

This point is strongly reinforced by our case studies – which typically involved the application of standardised, general-purpose tools – notably bulletin boards, email and, most recently, the Internet and the World Wide Web. Indeed, one significant observation across the SLIM investigations was that the most important social learning processes were not primarily concerned with the technical elaboration of the artefact[13] but revolved around the content and use of ICT. For example, the Danish Black-Out case study found little of interest in relation to the technology of the delivery system (in this case CD-ROM which is relatively stabilised). Learning instead focused upon 'the way to tell a story ... indicating a potential change in ... cultural genres' (Hansen 1998).

These features are rooted in the broad characteristics of innovation in contemporary ICTs. The research proposal for this study was developed in a period in which the nature of the delivery system/technical platform that would bring ICT into the home, the community and everyday life was still open. Whilst the Internet-capable personal computer presented a powerful exemplar, with very high rates of adoption in countries such as the USA, in other countries, for example Germany and to a lesser extent the UK, interactive television was seen as the basis for delivering commercial ICT services in the home. This focused attention on the battle for dominance between these two – with their very different characteristics in terms of the user interface and the capacity and potential uses of the 'back-channel'.[14] Such a vision has become less convincing, partly because of the continuing success and development of Internet and the TCP/IP standards on which it is based. The idea of head-on confrontation between digital TV and the Internet-capable PC has given way as players from both constituencies have sought to open up their offerings to other markets. The huge amount of information available on the Internet forces interactive TV players to present their offerings as a point of access to the Web, rather than an alternative source – even Microsoft was forced to abandon its plan to develop alternative information systems. Partly as a by-product of the hegemony of the Internet, the Internet standards and the Internet-capable PC have become the *de facto* standard for the delivery system. This is one of the reasons why the development of ICT applications has, in general, involved only modest levels of technical development activity in relation to the technology platform. This was certainly the case with the digital experiments investigated by our SLIM project. Most of the ICT applications surveyed have been configured from largely standard technological offerings – and, indeed, have tended to involve well-established applications of technologies that are already widely adopted, rather than particularly novel applications of 'high-end' technologies. ICT experiments and trials typically involved a personal computer with a CD-ROM reader or a network connection (mainly and increasingly the latter), combined with an assortment of packaged software components, often based on the emerging global *de facto* standards, particularly with the increasing dominance of Microsoft Windows operating systems. The important local innovative effort has been to knit together these different technical and cultural elements into specific ICT applications.

This has to been seen as reflecting the practical exigencies of digital experiments. There are two main reasons. First is the need to economise on costly innovation and learning. Many of the projects we looked at were relatively small scale, with limited budgets. The availability of generic ICT packages with facilities for easy customisation provides application developers with the option of minimising technological development costs and leaving shut certain technical 'black boxes'. For example, UK Bank developed its video-conferencing system by combining existing market solutions – configuring them together and customising the interface (Procter, Williams and Cashin 1999).[15]

There was some indication of a trade-off in the number of fronts on which innovation was pursued between innovation in core technologies and in applications. Where established technologies were used, time and energy could be conserved for experimentation around the development of content and usages. Conversely, projects with a stronger technical development focus were often those in which less attention was given to technology appropriation. A number of technically centred trials seemed to be conducted in line with the rhetorics of technology supply – which presented ICT artefacts as providing finished solutions to user requirements: experimentation around technology use was therefore not seen as possible let al.one beneficial. However, such technically driven visions were a frequent source of difficulty. This was very clear in relation to the educational technology projects, in which the presumption that ICT would provide a technical fix – e.g. as a means for teachers to obtain and deliver educational content – led to a failure to consider pedagogic models and concepts of the social context and processes of education and integrate these with the educational ICT applications (Slack 2000a, Buser and Rossel 2001). Of course, as we see later, even the most technically focused project also involved experimentation and learning in getting these to work effectively.

Perhaps in part as a result of these trade-offs, the digital experiments that we identified and studied in SLIM were often far removed from the visions of radical change espoused by the technology pundits and forecasters. Whilst technology commentators and suppliers have often articulated futuristic visions of impending social and technological transformation – suggesting that new technologies that are here or just around the corner will offer radically new ways of doing things – the actual offerings that have been put onto the market have in the main been rather more modest in scope, and have tended to be little more than incremental developments of existing products and services. When faced with a conflict between the attractions and uncertainties of the rhetorics of radical change, developers seemed most likely to resort to their existing knowledge, routines and market links. As a result many ICT applications were modest extensions to existing practices. Once the strident claims of the futuristic visions are put to one side, this observation may not be surprising – wise and realistic investors will seek early returns and avoid uncertainties by building upon their knowledge of and links to their existing markets. This is certainly the case for most of the projects we studied, which represent only rather modest innovations. We did, however, also encounter some attempts to restructure existing markets and established understandings. We will return, below, to this dilemma regarding the need to align with, or to transform, the requirements of existing consumers.

A concern to minimise the amount of technical innovation that experimenters have to embark upon also underpinned the second factor. This was the desire to ensure that systems are compatible with the existing technical infrastructures and practices of potential user communities, and the industry standards that are emerging here. There is little point in developing a sophisticated application if it cuts out many potential customers by its presumptions about what technical facilities and skills they possess. UK Bank found this out to its cost when it developed an on-line banking system that could only be accessed by customers with a particular and relatively recent version of a browser – and one that moreover required the user to have a relatively recent and powerful personal computer.

Some of the longer running projects we studied emerged prior to the global hegemony of the Internet. The project developers have found themselves more or less compelled by user expectations (over and above any enthusiasm they might have had for such a development) to upgrade their offerings from proprietary systems or low-end generic communication systems such as bulletin boards to web standards. This is exemplified in three of our community information/digital city case studies. The cases in point are the Amsterdam Digital City (DDS), the Copenhagen Base and the Scottish Craigmillar Community Information System.[16]

We thus find that the technical infrastructure of delivery systems for digital experiments are developed around global Internet standards and cheap standard commodified solutions. There were exceptions. The kiosk developed for accessing the Antwerp Digital City (DMA) was deliberately designed to regulate use (Steenhoudt 1997). The keyboard is too low for the average user and it is installed at a too acute an angle. This design improves access for the people in a wheelchair and makes it extremely uncomfortable for extended use by others. By inhibiting people from using the information kiosk for too long, especially for surfing on the Internet, more people can make use of the services (van Bastelaar and Lobet-Maris 1999)! Even here, however, the component technologies, albeit not the final physical configuration, were broadly standard.

More generally we found ICT applications being aligned to existing delivery systems. This is not surprising. Creating a new distribution channel is extremely expensive, and can only be attempted by the largest players and even then at great risk. Whilst some may see benefits from going it alone, the overwhelming majority will seek to offer their products into markets constituted by existing delivery systems. This is particularly relevant for the producers of cultural products. They can maximise their markets by opening up their offerings for different media channels. Thus the Virtual Museum of Colm Cille ported its products from electronic museum displays to a CD-ROM product and to a website.

We should not presume that the current alignment regarding the ICT delivery platform will be permanent. Indeed, the emphasis has now switched to huge expected markets for products based upon mobile telephony rather than Internet-enabled PCs. Mobile and wireless technologies offer very different service opportunities and very different affordances as information devices. The difficulties and uncertainties inherent in mapping out successful applications, and understanding how to port existing applications from PC to mobile are attracting considerable attention (not least amongst the firms which have invested heavily in 3G licences). Development in this area is indeed one of the most interesting and difficult, as bringing different platforms, business models and architectures together brings new complexity as it challenges basic assumption of who users are, and their relationship with various parts of the technology and content supply industry.

GENRES AND METAPHORS: APPROPRIATION EXPERIENCE AS A DESIGN RESOURCE

Thus far we have considered two sets of resources that constrain, enable and pattern design: knowledge and representations about potential users and existing technical components/practices. The third element we now consider comprises sets of ideas about how ICT artefacts may be used and useful which may have achieved some wider currency. We distinguish here between *metaphors*, relating to particular kinds of application (both at the level of specific functionalities/tools and the ways in which navigation opportunities may be presented) and *genres* in relation to more generic and widely applicable ideas about how a user may interact with an ICT product/service. Such common understandings may play a crucial role in social learning. They can help unite the perceptions and activities of the designers and consumers of a product. They can thus assist uptake of technologies as well as helping guide design. We will discuss here how this sort of generalised appropriation experience can help inform design. We return to this again in Chapter 6 in relation to the appropriation of artefacts.

If technologies were designed entirely anew then they would be truly opaque when they appeared before us – as perhaps the use and utility of modern technology might be wholly perplexing to an infant or a hypothetical representative of a pre-industrial society. The fact that our children have become socialised from their earliest years to the idea that the television and video etc. have buttons which when pushed make things happen may lead us to underestimate the taken –for granted knowledge that we, as members of a technological culture, share about how technologies may be used. Partly because these understandings evolve gradually, some of the most insightful studies here have been historical. Marvin (1988) shows how, in the nineteenth century the newly emerging technological systems of telephone and the electric light adopted and adapted the 'codes and grammars' of existing technologies.

These sets of ideas inform the work of designers and developers as well as guiding and stimulating users in their encounters with technology. Norman (1988) introduced the idea of *conventions* that operate as cultural constraints on design. We have borrowed from cultural studies the concept of genres (Hansen 1998) to analyse the role of widely accepted notions about how an ICT artefact may be used, and how they shape both the design and the appropriation of new technologies. Film studies emphasized the elaborate codes, grammars and rules of production developed by cinema and the mature ability of viewers to decode the film text. Similar processes were under way in relation to digital media content but were still at an early and weakly developed stage (Monaco 2000). Such genres serve as an important resource for designers (in reducing uncertainty about consumer acceptance) and for users (in terms of facilitating understanding of the uses and affordances of artefacts and thus their ease of uptake and usability). Established genres (in the sense of paradigms of design and use) can thus be a valuable focusing device: stabilising and harmonising designer and user understandings. Norman's (1990) concept of affordances refers to the perceived and actual properties of the thing, primarily those fundamental properties that determine just how the thing could possibly be used. To the extent that the perceived and actual properties/uses of an artefact can be aligned, then the artefact can become 'invisible' (Norman 1998).

On the other hand, continued technological innovation may undermine established understandings – particularly with the latest generations of ICT application which constitute new digital media, and which are presented as affording new kinds of data display and new kinds of interactivity between system and user, and which may involve the erosion of established genres. And if new genres are to be established, what are the processes that assist it. For example, having established one kind of usage can we extrapolate from this to another area? Can we generalise particular established usages (can we successfully use them as a metaphor for future design and use)? Crucial social learning processes are currently under way.

In relation to genres of using ICT applications and especially new digital media, we can begin to pick out the different elements around which design and other social learning processes are taking place. We can differentiate the form of the medium from its specific cultural content, and distinguish, within the former, between the tools used to obtain access and they way in which that content is organised. In this way we can distinguish, for example, between:[17]

1. tools and navigational devices (e.g. buttons, search facilities, scroll bars)
2. scripts, narrative architectures and narrative devices (e.g. relating to how stories are told, or to how a user may enter, make a journey through and leave a programme) and maps/navigational structures and coordinates (i.e. showing where a user is within such a journey)[18]
3. specific information and cultural content.

In relation to tools and navigational devices we point to the increasingly pervasive interfaces of the Internet and the World Wide Web, building as they did upon elements of the dominant WIMP (Windows, Icons, Menus, Pointing device) computer interface environment in personal computing. In terms of learning new genres, our project was particularly interested in scripts and narrative architectures (and 1) tools and navigational devices in so far as it impinges upon 2). We explore below how the Amsterdam Digital City

(DDS) deployed a geographical metaphor of a city (with squares and houses) to present and organise the spatial layout of its various 'pages' (van Lieshout 2001). The *Virtual Museum of Colm Cille* used a strikingly similar spatial layout presented through the metaphor of a museum with intersecting rooms and display walls. In this way we find a process of borrowing from areas of social life which are well understood (or, to be more specific, where understandings are well established and broadly shared by many sections of society) – and in particular, as we see below, a process of borrowing from other established media. This parallels Marvin's (1988) observation about adopting and adapting the 'codes and grammars' of existing technologies. There is, however, a tension between things being new and things remaining the same.

Much has been made, in discussions of digital media, of the rather different properties of this new medium compared with established physical media (e.g. printed paper, vinyl records) and its affordances in terms of new flexibilities in the way information is combined and accessed. These include '*multimedia-ness*' – the scope to present data through more vivid forms than simple text (i.e. voice, graphics, moving images) and greater *interactivity* between the user and the product or other users (e.g. Schmutzer 1999).[19] Expounding upon the latter, for example, one strand of technical enthusiast culture has emphasised the ability of electronic media to radically transform the relationship between producer and consumer of cultural products; of artist and audience (Qvortrup 1999). Some have called for an end to the conventional fixed and linear narrative structures of physical media. The cultural product is transformed from finished object to a resource for reworking by its consumer. This literally occurring with music sampling and remixing, and personal blogs and web pages, and is the basis of many video games where the cultural product is explored and experienced differently by each player, new graphics are supplied by users, and new virtual worlds created by players of multi-user games.

The Black Out study explored these themes. It suggested that many digital media products involve rather limited types of innovation/social learning. The first generation of products were driven by the desire of the commercial computer market 'to sell the media without any real regard for content or creative forms of interactivity'; the second generation of CD-ROM product was used by cultural industries to present their traditional offerings without development of any new media content. As a result, Hansen (1998) argues, the digital media products we have seen to date offer little more than re-cycling of old media content in new formats (e.g. CD-ROM etc.) with little real innovation in terms of media forms or 'interactivity'. Only a minority of products have fully exploited the potential interaction between media and content (Hansen 1998, Preston 1999).

Hansen (1998) sees the Black Out case as an important exception in this respect. Here the change is not only in the form of presentation, but is also reflected in the content of the story and structure of use.

Case Summary 5.3
Black Out – integrating interactivity into narrative forms
Black Out is a CD-ROM product developed in Denmark. The starting point for Black Out was on the content development side, with an author who wanted to tell a story in a digital form. His aim was to create a hybrid product combining the qualities of literary and computer game genres. Here digital media producers integrate interactivity into the story. They have used digital media to break the linear model of a story on traditional e.g. paper-based media, designed to be read sequentially from the start to the end. The original idea was that users should each be able to create their own story. Though this has is not been possible in Black Out, the users are able to vary elements of the story. No two users get the same story, when they play. They each get their own version with their own details. This creates a space for experiences of the story which reflects the personality and choices of the user. (Hansen, 1998)

The new kinds of interaction between user and artefact that may emerge with new digital media are shaped by the technical dimension of the delivery system. Kerr (1999) noted, in particular, that in their studies of cultural products, 'the distribution platform influenced the complexity of the content and the degree of interactivity between the final-users and the media and/or with the producers' (Kerr 1999: 46). The CD-ROM was designed as a more immersive game like space in which one explored and interacted with the media directly via the keyboard and mouse; in contrast, the website 'was viewed within the guiding context of a browser window'. Email facilities were built in to allow a degree of interactivity with the designers. She concludes that 'the form and content were influenced greatly by the technical limitations of the distribution media' (idem).

Of course, once content is available in digital form, it may readily be ported to different delivery systems with very different user interfaces and navigation modes. The Nerve Centre study developed cultural products for distribution on off-line media (namely CD-ROM) and an on-line media, the World Wide Web. We see here and elsewhere how the boundaries between on-line and off-line media are becoming blurred – with widely adopted products like the Encarta CD-ROM encyclopaedia acting as a point of access to websites.

Hansen (1998) reminds us that it may take some time before we can achieve a full understanding of the affordances of new technical capabilities, particularly if these differ significantly from existing media forms. On the other hand, some elements of ICT operation have very rapidly become globally communicated and accepted (the ideas of windows, buttons, the

World Wide Web metaphor). The point about such models and metaphors is that they are not only a resource for design, but also help to contribute to appropriation.

We should remember that the imputed 'digital revolution' (or transition to an information society) is still in its infancy. In particular, we are still in the early stages in the process of developing new *genres* (in the sense of literary genres – i.e. corresponding to the second type distinguished above) in relation to the narrative architectures and navigational structures for ICT and digital media products. We are referring here to the sets of protocols and ideas about cultural content products and how they may be 'read' that can unite and guide the activities of reader and developer alike; and to devices and narrative structures that inform players how to write narratives/products and how to read them. The history of cinema shows us how these types of convention emerged and evolved very gradually (McBride cited in Preston 1999). With hindsight many of today's digital media offerings will doubtless seem as quaint to the next generation as the first movies seem to us today. The appropriation of genres is thus crucial to the future of digital media design and ICT applications more generally, and we will return to this in Chapter 6.

Best practice exemplars and metaphors of design/use: insights from the digital city and community information cases

Given these considerations, successful examples of ICT adoption and use could become an important resource for designers and developers, as a source both of ideas about how to apply ICTs and of evidence of its acceptance and uptake. However, the resort to 'best practice exemplars' still presents problems, regarding both the selection of appropriate exemplars and the diagnosis and attribution of success. How successful was a recipe? What were the reasons for its success? What elements can be transferred to a new setting? Will they prove successful there? Our framework suggests that there may be a delicate interplay here between the benefits of alignment and the need for local differentiation, balancing the benefits of mutual reinforcement around established models and the need to cater for local specificities.

Our studies of digital cities and community information systems, by focusing upon a rather similar set of applications give us very interesting insights into these processes (van Bastelaar et al. 2000). We review these below. In addition, rather similar conclusions could be drawn from a set of studies of cybercafes in different parts of the UK and Ireland (presented in Chapter 6 as they primarily concern the appropriation of commodified supplier offerings).

For example, the Craigmillar Community Information Service (Case Summary 4.4) was explicitly initiated by a desire that Scotland should launch a project to emulate the Manchester Host – one of the earliest community information systems. The idea of using ICT to integrate communities drew explicitly on the 'electronic village halls' in the Manchester project – although the ambitions and scope of the Craigmillar project were considerably scaled down in terms of geographical coverage and scope (Slack, 2000b).

Subsequently Fredrikstad in Norway drew inspiration from the *Gemsis 2000* project in Salford: the Frihus 2000 feasibility study report, reviewed a number of international exemplars, and selected the Gemsis 2000 project in Salford as presenting a more *realistic* solution for a town such as Fredrikstad. Brosveet (1998), in his SLIM study of Frihus, concluded that Gemsis 2000 represented a *questionable* choice (given that it was an example of the development of a future urban ICT highway) as model for Fredrikstad, which was a rural area at the *periphery* of ongoing activities. UK observers might be surprised to find these models being taken up; however, external perceptions of the success of Gemsis 2000 might well have been much more favourable than that of players directly involved! This points to a more general risk that it may be hard to evaluate exemplars, particularly when they are far removed. What weight can be attached to success stories (and, of course, stories of success are more likely to be projected by those involved than stories of failure)? In such a situation, external observers may confuse the expectations and objectives of a project with actual achievements.

In the case of digital cities the diffusion of these exemplars was actively promoted, albeit in an informally organised manner. Thus the early experiments with Manchester Host provided one of the exemplars that informed the formation of the telecities network (an off-shoot of a European Community [EC] supported network of major European cities), allowing this experience to be presented more widely.

Subsequently, the Amsterdam Digital City (DDS) came to provide a widely known and highly convincing exemplar for such public information systems. In particular it offered an integrated view that encompassed ideas about the design and use of such information systems. In particular the analogy between the arrangement of web pages and the layout of a city into squares and houses etc. provided a compelling and readily understood metaphor about the design of the user interface and how the user would navigate through the information space. DDS has spawned a huge array of digital city initiatives – a testament to the success of the telecities initiative as a vehicle for disseminating these ideas, and to the attractiveness of the DDS design/use metaphor.

Once the concept of a digital city had emerged and come to be seen as effective in one place, there are obvious opportunities for its wider adoption. What is striking, however, when we consider some of the wide array of digital city initiatives that were studied by our project, is the diversity of outcomes they reveal from projects that apparently drew inspiration from the same sources: especially DDS.[20] For example, the Digital Metropolis Antwerp (DMA), has a very similar structure to the Amsterdam DDS project due to the collaboration between the two cities in its development. However, DMA utilises some rather different design metaphors such as a 'bridge' (between the administration and the inhabitants) and 'quarters' which indicate specific themes (sport, culture, education, etc.) rather than specific places or streets. DMA is more closely integrated with the local authority and geared towards integrating the citizen with one-line public services. So what was apparently a twin sister project became transformed in a different translation terrain.

The possibilities for diverse interpretation of a 'best practice' exemplar should not be seen only as a problem. Attempts to mimic particular best practices solutions in different contexts are likely to fail. Some space for experimentation was necessary for successful appropriation of a new technology. This was amply demonstrated by the Educational Technology cases, as we see below when discussing *unanticipated uses* of technologies.

These observations illustrate some of the key feature of successful metaphors. What makes metaphors such as 'The Digital City' effective is precisely their ability to provide meaning to diverse groups. And this seems to involve not only their immediacy and rhetorical compulsion but also their openness openness to local reinterpretation and redefinition. Thus the metaphor of the digital city has succeeded not merely because it provides a convincing description of what a new system might involve – but also offers a rather open and comprehensible frame that can readily be grasped, for example, both by designers and users of information pages and can cater for a variety of activities. Successful metaphors are not those which simply impose a uniform outcome, but rather those which allow a range of players to own and appropriate them, and redefine them according to their own particular circumstances. As such they are complex entities, nests of concepts (e.g. about user, technology, uses) rather than single ideas, which provide 'hooks' that may engage the interest of diverse audiences. As we note in Chapter 6, they function like 'boundary objects' in maintaining commonalities of meaning across dispersed social and geographical sites (Star and Griesemer 1989) – indeed, what makes this kind of metaphor for ICT design and use particularly important is its communication across large swathes of our social fabric.

STRATEGIC CHOICES AND DILEMMAS IN REPRESENTING USERS IN DESIGN: ALIGNING WITH OR TRANSFORMING USERS; PREFIGURING OR UNDERSPECIFYING USERS

It is now time to return to the issues surrounding the representation of the user in design. We have come a long way from the naïve conceptions introduced at the start of this chapter of design–representation as a simple reflection of user requirements – derived perhaps from the inductive collation of an array of actual end-users. The preceding discussions pointed instead to the complexity of the representation–design process in the development of ICT applications. We have come to see design as a rather complex 'configuration' process that is strongly contexted, in the sense of being patterned and channelled by its particular historical setting – for example by the sets of specific resources and constraints (including, for example, the availability of component technologies; various kinds of knowledge of existing uses and users). Our range of case studies thus highlights the diversity of processes and outcomes depending upon the design and development arena and its insertion into a specific translation terrain. We found rather marked differences even for rather similar types of initiative, as exemplified by the range digital city cases we studied. These features of context and process bear upon the ways in which representations of the user and use are approached. It is hardly surprising therefore that there was no single approach to user representation across our cases. Across the range of cases we could identify some of the strategic choices that actors made in grappling with their particular context. The dimensions of these choices highlight a number of dilemmas and contradictory considerations.

We will now explore two of the key dilemmas regarding the development of representations of the user and context of use and their embedding in artefactual design:

1. Whether in catering for future possible users, to seek to align with or conversely to transform existing user models and practices.
2. How far it is necessary or desirable to build specific models of and references to particular users into the designed artefact – or whether to leave the artefact as a generic solution.

Building on existing users/markets or transforming them?

We have already briefly commented on the discrepancy between the revolutionary rhetorics of technology supply, with their emphasis on transforming, reinventing the user or creating them anew, and the more conservative patterns of incremental change that prevailed across most of our case studies in which expectations about the user and user setting were broadly aligned with existing practices and understandings. This highlights the difficulties in trying build new markets from scratch or to transform existing markets (particularly in relation to goods targeted at mass consumer markets) regarding the need to relate to the established understandings and requirements of existing consumers. To 'sell' a new product to someone it is necessary to convey to them some sense of its utility and how it can be used. It may be difficult to do this in relation to wholly novel ICT applications. Instead, it may be more easy to convey the utility to users where the design and the promotion of the artefact can borrow from or refer to existing understandings and experience. The importance of conveying such a common understanding is underlined when we consider that consumption of an artefact by a user is not only about its functional properties, but also about its perceived affordances and relevance to the user, and, as we noted in Chapter 3, this is also a question of the (re)production of meaning and identity (Sørensen 1994a, 1996). There was thus an ongoing issue across all the cases we studied about whether the design of ICT applications should be constructed around existing understandings of current groups of 'users' and market segments or around visions of how these user markets could be transformed. We are able to explore this in detail in relation to one widespread parameter – regarding the gender construction of ICT artefacts. We found a number of cases in which particular attempts had been made to transform existing gendered market boundaries and user characterisations. These investigations therefore afforded particular insight.

Insights from the gender construction of artefacts

An insightful illustration of the ways in which technologies have tended to be designed around existing socio-economic and cultural parameters, and of developments that may serve to transform these, can be obtained by examining the gender construction of ICT applications.

Our studies confirmed a theme in much contemporary literature on gender and technology that computer-based technologies have traditionally tended to be constructed as predominantly masculine artefacts and that their domestication has generally tended to follow clearly gendered patterns. As in other technical spheres, computer competencies and enthusiasms are heavily gendered as they are deeply embedded within, and shaped by existing social and cultural conditioning. As one case study report puts it: 'males are the keepers of the knowledge of the machine, they control the practical use of it, and through its placing and signification, it is symbolically given meaning as their "toy"' (Spilker and Sørensen 2002). Digital media content styles might also be deemed to reflect the user base at present which tends to be male, urban, well-paid, college educated, 25–35 years old etc. (Preston 1999).

However, our studies also emphasised the dynamism of these situations. They add weight to recent discussions of the *mutual shaping* and reshaping of gender relations and technology (Berg and Aune 1994). Most of the initiatives we encountered tended to operate within rather gender stereotyped parameters; we further we found that males have tended to predominate in the production teams involved in the construction of ICT applications (Preston 1999). However, our set case studies also include some attempts to move outside these parameters. In particular, certain innovative initiatives have sought to exploit the potentially large but undeveloped women's market for ICT products and services (it may not be a coincidence that these were both in Norway, a country in which feminist concerns have become institutionalised, for example in public policy).

Perhaps the most striking example was Girls-ROM (see Case Summary 5.4; Spilker and Sørensen 2000, 2002). We could interpret this as a **socially** radical project, which we could see as an attempt to transform the gender identification of the technology and to create a new market. It involved an attempt to open up an established medium/delivery system (PCs with CD-ROM readers) to a new market – of young women. Despite this demonstrably radical social goal, what is interesting about the project is the rather pragmatic approach it involved. Girls ROM, at its launch, was built upon fairly conventional gendered understandings of this target market – and extrapolated from existing cultural product markets (e.g. girls' magazines) and included, for example, somewhat stereotypically gendered features such as a 'secret diary', which it deployed within a new delivery system. This **incremental model** of change (in terms of working within established parameters both of an existing market segment [girls' magazines] and of an established delivery systems [CD-ROM PC]) offered a more realistic and realisable pathway to the wider appropriation of digital media products than might have been possible with a 'Big-Bang' approach – a head-on attempt to achieve rapid and profound realignments of both technology and markets (Spilker and Sørensen 2000). Similarly the subsequent launch of HomeNet sought to build upon rather than transform gendered identities. And even the emerging sub-culture of young women computer enthusiasts identified in our 'Spice Girls' study (Case Summary 6.3), though becoming deeply involved in an area hitherto seen as the exclusive province of young 'techie' men, emphasised their gender normality.

Case Summary 5.4
Girls ROM: constructing content around gendered cultural forms
This was an initiative by a media/technology SME to construct digital cultural content around orientations/interests of young women, through a CD-ROM

that was given away with a magazine (*Det Nye*) oriented towards young women.

The CD-ROM included elements designed to be attractive to young women – e.g. a private diary function. It thus aligned itself with existing cultural forms in conventional media for young women. *Radical change* in the gendering of ICT markets was thus pursued through an *incremental strategy* – bringing together two established markets (magazines for young women; personal computers with CD drives).

The Girls ROM project has had longer-term effects: though driven by a media/technology SME, the project involved a network of actors, including advertising interests and a large magazine publisher. Since then, the latter has launched its own larger-scale HomeNet project in collaboration with other Norwegian publishers. This too seeks to address women users as a specific market segment within a broader service that is oriented to different (gender, age, etc.) categories of Norwegian home users. The study seems to indicate a rapidly growing awareness and 'social learning' on the part of the mainstream media corporations and advertisers that it is in their interests to construct the widest possible audience/user base. This is prompting them to expand the range and forms of services which address specific gender group identities and thus address related aspects of 'exclusion' – at least from the more commercial or consumerist digital content products (Spilker and Sørensen 2000, 2002).

These discussions about the gendering of ICT bring us back to the need to address design in the context of broader social learning around technology appropriation and domestication, and the interplay between these. Sometimes these interactions may serve to stabilise technological forms; at other times – and it would appear today – they may reinforce changes. Our cases added to the growing body of evidence that we are on the threshold of a sea change in the gender identification of ICTs. The growth of the Internet and the stabilisation of certain tools and functions (especially email, which continues to be the most significant application for most users, and the Web) has consolidated the role of the computer as a device for communication and sociability, and thus a potentially more convivial device for women, and as a useful medium for productive professional and social activities rather than a 'toy' for technically oriented men to play with. We can distinguish two elements here; on the one hand, the openness openness of these generic capabilities – increasing their potential attractiveness to many different kinds of user; on the other, some explicit moves to make digital media content more relevant for women. The culture of computing is also undergoing change, as is evinced by the emergence of sub-cultures of computer enthusiast young women explored in the SLIM study 'From Spice Girls to Cyber Girls' (Nordli 2001). Very rapid rates of uptake of the Internet has led to a situation in which in the USA and (especially Northern) European countries, the existing markedly higher levels of access to the Internet of men over women, has gradually become eroded, so that now gender is a less important differentiator of access than other demographic variables such as income and in particular age (Sørensen and Stewart 2002). This is not to suggest that ICT has somehow become gender neutral. There continue to be clear gender differences in the patterns of usage in terms of the kinds of activities undertaken and the duration of usage.

The Internet in particular has opened up more imaginative possibilities for technology, moving it away from its historical associations with technical specialists and a narrow, wealthy and highly educated elite. The diversification of computing applications has increased the potential relevance of these technologies to different groups in society. And we also see explicit initiatives to encourage wider access to ICTs in general and also to overcome particular forms of social exclusion both pre-existing social barriers and – as evinced by discourses of the digital divide – those which may arise with the adoption of ICTs (Sørensen and Stewart 2002). We find a range of attempts to increase the perceived relevance of technologies to different social groups as well as women. For example, the Craigmillar Community Information System, in its concern to increase its local user-base and thus its attractiveness to external public funders, fostered a number of specific user groups to extend the size and range of users of its services within the local community, including a 'cybergrannies' group (see Case Summary 4.4). As in the case of 'women's' technologies these groups of users/uses were built around stereotyped constructions of user needs (in this case the characterisation of older users' needs as a way of recording their memories) (Slack and Williams 2000). The same dilemma thus appears regarding the enrolment of potential users/customers. On the one hand, the supplier seeks to demonstrate novelty and difference; on the other hand, the design and promotion of the new product needs to draw upon existing understandings to impart meaning and validate adoption. In the case of 'Local Ireland' the emphasis was upon adding value to existing content from traditional media while distributing it more widely through the Internet (see Case Summary 5.5). This conservatism reflects the costs and difficulties of producing novel high production-value content and building new technology infrastructures – and the exigencies of creating new markets.[21] We will come back to this point in considering processes of appropriation and domestication processes in Chapter 6.

Case Summary 5.5
Local Ireland: local communities and local culture on the Internet

This case study examines the development of an on-line project called 'Local Ireland' – a set of websites, primarily text based, with accompanying communication services – led by a new firm which subsequently developed close links with the national PTO, Telecom Eireann. The project combines a global technology with the development of local content and skills; it involves the development of voluntary cooperatives in each county in Ireland to provide the content at a local level and train local people in the necessary inputting and IT skills. At a higher national and regional level the initiator company itself or various sub-contract actors provide content.

The project aimed for the innovative use of technology rather than innovative content development or novel kinds of interactivity. It aims to facilitate local content provision and repackage (add value to) existing content from traditional media while distributing this to the widest possible audience over the Internet. The concept is driven by marketing considerations even as it adopts much 'community' orientated rhetoric. A key underlying idea is to develop a heavily promoted/marketed Website which provides information relating to Ireland (at a national, regional and local or parish level). This strategy seeks to attract users of the local/community information services and at the same time build a captive audience for the key actors to promote/sell subsequent commercial service operations. In addition, the case provides a study of an entrepreneurial company which has exploited low barriers to market entry, particular promotional tactics and marketing strategies and networking possibilities. Though the outcomes were eventually highly successful, the case also highlights difficulties (and consequent delays) in obtaining funding and in developing the necessary software and infrastructures before the project can be effectively launched at local levels.

HOW MUCH DESIGN/DEVELOPMENT?

Closely related to these considerations about whether to align new offerings with existing understandings and practices, questions also arise about how far to prefigure the intended future user. Should ICT application development attempt to build particular concepts of uses, users and contexts of use into an application, as opposed to leaving these questions relatively unconstrained? Underpinning this choice are competing views of the designed artefact, on the one hand, as a finished 'solution' to particular social needs/user requirements, and on the other hand, as inevitably unfinished in relation to 'needs' and requirements that are more diverse and evolving more rapidly than can be adequately captured by design. The latter view sees prior *design* (and earlier cycles of design and appropriation) *as a resource for social learning* – most immediately through the appropriation activities of intermediate and final-users as well as in future development cycles. As we argue in Chapter 6, most of the ICT applications we studied were, in this sense, unfinished. Indeed, according to the social learning perspective, this principle is applicable to all forms of ICT. Artefactual design is always to some extent generic in relation to specific users – requiring work to be done to locate it within particular local socio-technical contexts. However, there are different levels of design generality, reflecting distinct strategies for technology supply/acquisition.

Technology supply/acquisition strategies

Suppliers/developers must develop some strategy relation to their intended market – in terms of how they attempt to anticipate future users' requirements and to provide (or at least present) offerings that correspond in some ways to the users' technology acquisition strategies. When we come to consider the strategic choices developers faced we encounter rather similar dilemmas to those just discussed about the enrolment of users.

Consider on the one hand, the arguments for building a system around specific user representations and their articulated requirements. The supplier needs to convince potential users – and above all those responsible for purchasing decisions – that their offerings will meet their particular purposes and will be sufficiently beneficial to justify their acquisition. Highly dedicated applications – i.e. closely configured around a specific set of articulated requirements – may have the advantage of increasing the utility of the application to particular users (where these match the particular requirements and priorities of the users in question) as well as increasing the *perceived relevance/utility*, in terms of making these inscribed utilities and their advantages more apparent to potential future users.

On the other hand, such dedicated applications run a greater risk:

1. of failing to capture accurately the current requirements of target users;
2. of excluding other potential users whose circumstances and requirements are somewhat different;
3. of foreclosing (rather than enabling) the active role of users in reinventing the technology and its uses in the course of its appropriation.

There may be an important temporal element here. In the short term it may be more feasible to enrol potential customers to buy-in to applications which are closer to their immediate perceptions of their current needs and how technology might fulfil them. However, if this incorrectly anticipates the kinds of use that will ultimately prevail, it may deter potential users and constrain their ability to realise their objectives in future. We see this in the problem of organisational 'legacy systems' which embed in software past organisational structures and business practices. The often observed difference between early and late adopters (for example in their skills and attitudes to technology [Rogers 1983]) may be one factor behind the emergence of baroque technologies that are more complex to operate than typical consumers feel comfortable with (Norman 1988).

There is thus a continuing dilemma surrounding the design and development of ICT products and services about how far to prefigure the user in systems development and build specific user representations into designed artefacts. There is a danger that facilities that are too abstract run the risk of failing to engage and convince potential customers. Building applications around specific concepts of user and use may increase their perceived value to particular users.

However, the above considerations suggest that there are risks in seeking to foreclose user choice too tightly, and in trying to move too far beyond current practices and models. These factors may mandate in favour of adopting more generic design approaches. This may involves active strategies to build upon successful specific applications, but 'decontextualise' them – to **design out** from the artefact **reference to its specific contexts of origination and use** which may limit its future use and market – or more precisely to **'re-design'** and re-present the artefact to make it more generic and open it up to broader markets. Here we note again the preponderance, revealed in the ICT applications we studied, of generic information tools (email, bulletin board, the World Wide Web).

We have already drawn attention to the more general point that artefactual design is inevitably generic to some degree in relation to specific users. Since not all users can be directly involved in design – even where user involvement takes place, selected users must inevitably to some extent stand proxy for their peers and for future potential users. Indeed, successful system design depends on an ability not just to capture the specificities of the user context, but also to translate these into a form in which they can be more widely used. Whilst the design fallacy conceives the improvement of design in terms of building-in ever more knowledge about users into the artefact, there are also risks in trying to prefigure too closely the user and their purposes and in seeking to foreclose user choice around the expressed preferences of particular sets of users. There are issues around the building of representations of the user. Moreover, design is subject to a number of contradictory paradoxes – between making a solution specific and generic; between aligning with and moving beyond current practices and models.

These factors may mandate in favour of adopting more generic design approaches. We thus see strategies to build upon successful specific applications, but to **design out** from the artefact **reference to its specific contexts of origination and use** which might limit its future use and market – or more precisely to **'re-design'** and re-present the artefact to make it more generic and open it up to broader markets. Designers may need to balance between building solutions that are very tightly configured around particular local requirements – which may for example act as a barrier to utility and use in other contexts – and keeping the system more flexible. Schumm and Kocyba (1997) have described the related processes as involving, on the one hand, *decontextualisation* of this knowledge (selection and synthesis of specific requirements, its abstraction and separation from particular contexts, its codification to make it more widely applicable) and of *recontextualisation* (to implement this generic knowledge within particular artefacts). This involves a shift in perspectives from particular users to generic users – a translation from one to the other. For example, in the creation of an integrated ICT system for an organisation we may find a shift from specific organisational users to a generic user – and in the case of mass-market products, a shift from a focus on specific proxy users to an abstraction 'the user' that may be characterised differently (for example in terms of conventional market research conceptions of 'market segment').

These choices in modes of design (including the dichotomy we draw in the next chapter between an emphasis on accountability and on creativity) have an obvious linkage with the strategic choices we identified in Chapter 4 regarding the organisation of social learning. We find a range of different strategies for addressing this design dilemma, between at one extreme the *prior design* model – e.g. user-centred design, which seeks to build up a comprehensive knowledge of the user setting and purposes into the designed artefact by studying or involving particular users in artefact development, and at the other extreme the *laissez faire* model, in which the user configures standard generic components to their particular needs. Although many elements of design are divorced from users in the latter model (i.e. the design of component technologies), when we considered the whole product life cycle we argued that this represents a rather clearly 'user-led' strategy in the scope afforded to intermediate and final-users to actively select and configure elements to their particular purposes. This model has also been the basis for some of the most widely adopted ICT innovations – indeed, the most dynamic field of contemporary technology supply and technology appropriation – the World Wide Web/Internet largely corresponds to this model. An important feature here seems to be the existence of a sub-culture that shares a view of the usage of technology – particularly where this view turns out to be one that can be generalised to other groups (as we argued, for example, in relation to the success of evolutionary innovation model found in the Amsterdam Digital City initiative). A corollary of this it would seem, from the success of technologies such as the World Wide Web, that a particularly flexible supply strategy may well be to keep supplier offerings generic and provide for the active role of local appropriation players in configuring and adapting them to their purposes. Here the technology delivery system and infrastructure becomes a generic platform on which a wide range of content can be mounted – in some cases it becomes a medium, largely independent of the content being launched; in others it becomes a rather open communication channel (for example email).

These issues bring us to the concern of the next chapter, with the appropriation of new technologies. We started out this chapter with a criticism of the view of design as reflecting a single set of values. In practice it seems that ICT artefacts rarely embody a single 'script', but instead are built around a multiplicity of overlapping possible scripts rooted in the historical patterns of past cycles of technology design/use and the particular socio-technical arena, linking an array of actors with diverse commitments and experience.

NOTES

1. Mackay et al. (2000: 18 ff) review the various formulations that Woolgar has advanced for configuring which include 'defining the identity of future users and setting constraints upon their likely future actions' (Woolgar 1991: 59 note 17).
2. Since, as we note elsewhere, these studies of 'designers configuring the user' do not in general address both design and implementation, the impact of design choices on the user is largely imputed.
3. As Mackay et al. (2000) note, working within, but seeking to adapt and extend, the Woolgarian framework, configuration is not a one-way process; though designers do configure users, they are in turn configured by both users and their own organisations.
4. Though this point goes somewhat beyond the scope of the current chapter, it can be observed that there are obvious limitations to the role of ethnography as a method for requirements capture; it is an expensive and slow method of data capture. More crucially, effective requirements capture and design is much more than the accumulation of ever more knowledge about diverse user requirements; you could never do enough ethnography to create an all-inclusive account of, for example, a large organisation; nor could you simply induce a design solution from such a knowledge base. Effective design is not an induction process; it also requires the generalisation of certain selected elements based on some kind of accommodation, satisficing and prioritising between the specific requirements and preferences of multiple individuals and groups. The strength of ethnographic method is in picking up and providing insights into the intricacy of work contexts. However, in most circumstances its role can only be as a resource to deepen other methods of requirements analysis rather that as a primary requirements capture methodology.
5. Users may find additional benefits from adapting to standard offerings – for example regarding the greater availability of skills to maintain and use a package, or a desire to align to industry norms or best practice standards.
6. Indeed, one current in the social learning perspective talks of the co-evolution of technology and society.
7. As well as the representation of the users, an important influence in relation to the digital city initiatives and the design of the services was the way the city was represented. One significant difference concerned whether the city was employed exclusively as a metaphor (as in DDS – Amsterdam), or whether there is a representation of a grounded digital city based on a concrete city and its residents (as in CB). This raises questions about where the city limits of a digital city are and who may represent it.

8 This may be one reason why the adoption of domestic ICTs has often fallen far short of expectations (Thomas and Miles 1990).
9 As an aside we would suggest here that the defining characteristic of a radical innovation is the extent to which it cuts across existing institutional forms, in contrast to the earlier literature which tended to see 'radicalness' simply in terms of the technical properties of a new technology (see, for example, Freeman 1974).
10 As Birgit Jaeger has observed, the social learning perspective problematises naïve concepts of representative 'independent' users. Concrete users involved in panels or trials are transformed by their engagement with a technology in the social learning process. Judging the representativeness of concrete users and their social learning outcomes cannot be resolved by simple statistical processes but need to be carefully considered.
11 Despite any similarity of nomenclature, this usage of configuration is wholly independent and very different from the discussions of configuring the user. Our usage is closest to the third definition in the *Oxford English Dictionary*: 3. *To fashion by combination and arrangement*; as well as the more specific meaning in relation to Computing. 4. *To choose or design a configuration for; to combine (a program or device) with other elements to perform a certain task or provide a certain capability.*
12 Dan Shapiro and co-workers have deployed Strauss's concept of *bricolage* to describe this process of building what we describe as configurational technologies from largely standard tools (Büscher et al. 2001). Confusingly, Ciborra (1996) has simultaneously applied the same term to refer to the playful use of artefacts, tinkering with technologies and the consequent drift in the use of computer systems – a usage that is much closer to our concept of domestication.
13 This is not to suggest that 'the technical' was unimportant – indeed, as we see below, technical problems in configuring a system continue to represent a major barrier to ICT uptake and use. Moreover, particular delivery systems have their own affordances and technical limitations.
14 Linked to this view was an expectation that the design of the delivery system/domestic platform would be built around particular kinds of application/service and thus favour some kinds of ICT use over others. Indeed, clear moral stereotypes were articulated, for example between the liberatory uses of the Web and the commercial consumption mode of use of web TV (as captured by visions of ordering pizza with your TV remote control).
15 The backbone was provided by a commercial Computer-Supported Cooperative Work package which was configured to allow staff to communicate by telephone and video-conferencing while jointly inspecting and manipulating information held on other, industry standards, spreadsheet and word-processing software. The Bank's developers configured these elements together, customised the layout and switched off certain functions that were not seen as necessary.
16 Paradoxically, the widespread uptake of Internet and web technologies eroded the distinctive position of DDS and CCIS as information providers for their communities. Amsterdam Digital City, no longer an 'obligatory passage point', found itself in competition with an array of other providers, including many commercial services. Users were tempted away in the face of a much wider choice of information sources and ultimately the DDS information service closed. All that remains of DDS today is its role as an accessible Internet service provider.
17 Though different writers have used different terminologies, and there may be overlaps and interactions between these elements.

18 We are intrigued to note how terms have been applied from other media forms which may be textual [scripts] or graphical [maps].

19 Thus Schmutzer (1999) suggested that the interactivity of a medium should instead be seen in terms of two dimensions of user control:
   (a) control over the content of communication – in terms of the extent to which the user can influence the content exchanged in a communication process; and,
   (b) control over the flow of communication – in terms what influence the user has over which contents are exchanged when and in which order.

20 Jaeger and Qvortrup (1991) had made a similar observation earlier in relation to community telecottages.

21 Despite the rhetorics of the digital revolution, which predicted that new media would displace the old, studies of digital media highlight the extent to which new media reinforce, rather than replace existing media.

6. Social Learning in Technology Appropriation: Innofusion and Domestication

ISSUES IN INVESTIGATING APPROPRIATION AND DOMESTICATION

Let us start by briefly reflecting upon the issues that may arise in attempts to investigate empirically the processes of appropriation and specifically innofusion and domestication of technologies. The concept of appropriation refers to an array of often highly dispersed processes, occurring often over extended periods and across an array of social spaces and involving an important informal element.[1] It thus presents particular difficulties for systematic research, which we should perhaps consider before we go on to address the specific findings of our study. One established research method involves intensive longitudinal ethnographic studies. However, these have, in consequence typically been limited in focus, for example, to a very small sample of selected families (Silverstone 1991, Silverstone and Hirsch 1992).

As we noted in discussing our research design in Chapter 2, attempts to conduct an in-depth study of both the development and appropriation of ICT face considerable practical and methodological difficulties, not least because the time needed for a new ICT project to go from product design and development to its eventual implementation and use can far exceed the typical durations of externally funded research projects. The SLIM research centres adopted various strategies in confronting this dilemma, including utilising prior research within and outside the network and combining contemporary with historical analysis. The result was a rich set of integrated studies which allow varying insights into appropriation and domestication processes in different settings. Some studies were of projects which remained mainly at the stage of project development and did not provide an opportunity to address appropriation. It is important to bear in mind these differential opportunities in considering our findings.

Four of the studies in which we are able to examine final appropriation (from our array of some 30 investigations[2]) comprise cases in which the final appropriation is taking place within the same 'configuring constituency' that developed the ICT application. The application was both put together and used within an organised and bounded array of players (hence our use of the term 'constituency' rather than 'constellation' here, explicitly to flag this fact) – whether this was a single organisation deploying ICT for its purposes (Edipresse, UK Bank[3]), or a coalition of organisations (Cable School). The Language Course case also involved proxy external users (teachers and students) bought in in a rather tokenistic manner so that the pilot could be tested and demonstrated at a forthcoming trade exhibition (Mourik, 2001). Where appropriation is within a single organisation or close alliance of organisations, appropriation processes can be expected to arise more quickly – as the linkages between players involved are direct and appropriation is more readily subject to purposive management and incentives – than, for example, in mass consumption products, diffused through the market (where links will be indirect, and where the consumer may be exhorted to take part but cannot really be directed).

In contrast, mass-consumer products and services tend to have a particularly long development cycles. If we wish to investigate ICT application development, in the lifetime of a relatively short-term study, projects are likely still to be at the developmental stage, and not to have 'rolled-out' for wider use – thereby offering relatively little scope for studying final appropriation. The SLIM research programme investigated a number of products that were geared towards such mass markets and domestic consumption.

The time needed to set up and conduct studies is not the only constraint in exploring appropriation in such settings. The other, as discussed by Silverstone (1991), is the private and closed character of the domestic space. These barriers, of time, space and openness posed particular problems of access to understanding local appropriation (barriers which of course apply to technology developers as well as to social science researchers).[4] For example, there was little scope for examining local appropriation processes in the Norwegian Girls-ROM case – in which a CD was given away in a magazine (especially since the project remained something of a one-off). The final appropriation process here was rather ephemeral.

However, these constraints did not prevent us from examining appropriation. Some of the ICT applications were developed for use in and by commercial organisations; most of the educational ICT developments were geared around their experimental use within educational establishments. Many of the products developed for mass markets (e.g. in edutainment) were not exclusively for consumption by the individual in the privacy of his/her home. For example, cybercafes emerged as important sites for the collective and public consumption of various ICT products. Here, and in other places, appropriation was actively organised. For instance, following development of the prototype 'Virtual Museum of Saint Colm Cille' on-line display, the developers, Nerve Centre, embarked upon an explicit appropriation phase as they sought to encourage the uptake of their products.

Case Summary 6.1
Virtual Museum of Saint Colm Cille: learning by doing
The project was established within the 'Nerve Centre' project based in Derry City. The virtual museum exemplifies the process of *learning by doing* in that the 'Nerve Centre' had substantial experience in 'traditional' media such as film and photography, but this was one of the first projects undertaken using digital media. Further, the development of local content about the saint's life was undertaken with an eye to global markets and representations – dialogue between the local and the global both in terms of modes of representation and technology was integral to the development of the project. In other words local content, taken-for-granted representational conventions and technologies together with established practices had to be replaced by a praxis-based social learning in the transition to digital media.

The content itself was developed by and for young people, but other actors were enrolled in the project by the Nerve Center in order to gain funds and to 'ratify' the project through the investment of cultural capital by older citizens who had little idea of the potentialities of ICT. One concern was with the representation of a local area or culture to the world through the use of digital media. The expectation was that younger citizens would become increasingly aware of their local culture (and the value of it) through placing it into a global context. Further, we note what Kerr (2000, 2002) calls 'social learning through Trojan horses': through an attempt to enfranchise older citizens in the uses of ICTs by the use of local content using new media technologies expressing community memory and history.

THE DYNAMICS OF ICT APPROPRIATION OF ICT APPLICATIONS: PRODUCTS, SERVICES AND DIGITAL MEDIA CONTENT

Most discourses about the technologies of the emerging Information society focus upon the development of advanced ICT artefacts. In contrast, our study of social learning in multimedia has highlighted the importance of *appropriation* as a crucial stage of innovation and learning in the overall 'circuit of technology' from design to use (Cockburn and Furst-Dilic 1994). This perspective highlights the importance of *domestication* and the incorporation of artefacts in local practice and culture. In relation to contemporary ICT applications our analysis draws attention to the importance of *digital media content* rather than artefactual design

Some of the key potential difficulties and uncertainties surrounding the future of ICT applications are thus associated with technology appropriation rather than its design and development. The appropriation process is highly unpredictable and rather difficult to coordinate, involving as it does, large numbers of diverse players, many of whom may not be committed to the technology, but who may need to be *enticed* to use it. This issue is particularly important with products geared to a mass of private consumers who cannot be directed or managed in a 'command and control' model in a way that might have been possible with earlier waves of ICT implementation in the workplace. How, for a start, can potential customers be encouraged to see these 'strange' offerings as potentially relevant and useful to them? If it is difficult to get potential customers to commit time and resources to investigate whether these relatively unknown products are worthwhile, how can suppliers motivate them to make the more substantial investments of time and resources in acquiring the technology and learning how to use it? How, too, are people expected to interpret the different visions of technology, and make decisions between competing technologies and visions? Following on from this, the appropriation of ICT applications can be extremely slow – as we shall see later, it tends to operate on far longer time frames than initial technology design and development – and be rather obdurate.

The difficulties associated with appropriation were brought home sharply in our study of the multinational marketing efforts of Compuflex. Compuflex found that simply translating its US products into European languages was not sufficient for it to be able to market these products successfully in Europe. Other kinds of translation in underlying concepts of use were also needed to make them attractive. Even an organisation with the global financial and technical resources of Compuflex Corporation was unable to impose its offerings onto the market.

Case Summary 6.2
The Case of Compuflex: global technology meets local content and cultures

This is the story of how Compuflex, a global ICT supplier, sought to move from its established base in the production of task-oriented software (and some experience in offline CD-ROMs) into on-line ICT services and digital media content provision. It is a story rich in lessons about transnational social learning and about culture and content.

The case study focused on a small team of technical and content specialists in Dublin, charged with the localisation of on-line digital media content developed in the USA and destined for a number of different European linguistic markets; mainly in entertainment and information services. However, localisation proved to require much more than simple linguistic translation to make this content relevant and attractive in different European markets. The case study provides insights into the concept of 'globalisation' in relation to the cultural industries. It highlights: tensions and problems with the concept of 'global content', the importance of cultural specificities, and barriers to the consumption of foreign produced content.

The original Compuflex digital media content project can be defined as a clear 'failure' when judged against the corporation's initial aims, objectives and expectations. 'Compuflex' launched its new digital media content product

'CFN' in 1995 and by 1997, claimed to have 1.5 million subscribers in the USA, and another million worldwide. But in January 1998, just two and a half years after launch, CFN was redesigned and repositioned for a third time. The new strategy involved a move away from content (news and entertainment type programming) towards on-line transaction orientated services (Kerr 2000, Preston and Kerr 2001, Kerr 2002).

In analysing the domestication of ICT products and services we emphasise the complexity of the process involved in the development of meanings and practices of use around the technology, involving diverse users and intermediaries as well as designers and developers. We have already examined under the heading design/representation the efforts of designers to articulate concepts and means of use and convey these to users. We now address the role of users in selectively taking up (or rejecting), adapting and reinventing these concepts.

The interaction between supplier efforts and appropriation has complex and contradictory outcomes. Whilst noting some areas of dynamism in relation to the uptake of ICT products and services, developments remain very uneven – reflecting the various exigencies of the social learning process. In contrast to the compelling supply-driven visions of a rapid uptake and global convergence of cultural forms around offerings from ICT supply, our case studies highlight the continued importance of 'the local', and the intricate interplay between localising and universalising forces. This was reflected in a complex set of outcomes in the cases examined between local and global elements and between what was generalised and what remained specific. For example, there had been widespread adoption worldwide of certain discrete ICT tools and component technologies. However, this has not occurred uniformly across the board, and in particular it does not seems to apply in relation to more complex information and cultural offerings. This cultural conservatism, and emphasis on the local is, Preston (1999) suggests, particularly significant in relation to heavily cultural or symbolically laden content types/genres of digital information and content. It is in relation to this type of products (for example those concerned with local community history and community development) in which questions of identity and self-presentation are likely to be salient, where local knowledge and authenticity are likely to be crucial.

Despite the widespread expectations of the rapid take-off of new ICT markets, the social learning perspective, and the question of appropriation in particular, suggest that the extension of use of new ICT products and services may be rather slow overall. We note that the uptake of new technologies has often fallen substantially behind expectations, particularly in relation to large-scale technological systems such as new ICT delivery systems (for example the failure of UK Prestel [Schneider et al. 1991] and, to date, interactive TV).

However, we also note the *unevenness and fluidity of development* – and consequent difficulties in predicting the future development and uptake of technologies. We must also take into account historical experiences which demonstrate that, in some circumstances, technology can be taken up far more quickly and completely than anticipated at the outset. In this process, the significance and use of artefacts is often transformed from that originally anticipated. For example, fax took off very rapidly in the 1980s once the technology became cheap, robust and easy –to use (Coopersmith 1993). A similar story can be told of mobile telephony in the 1990s. The even more explosive growth worldwide in the Internet in the period of this study provides the most striking illustration of this point. The intuitively simple interfaces of the World Wide Web have become widely accepted. The latest, and perhaps fastest, example of 'hyperfast' uptake is provided by the mobile telephony Short-Message-Service (International Telecommunications Union no date). In none of the cases was the rapid growth in uptake recognised in advance. We can conclude that progress in adoption of ICTs is likely to continue to be rather uneven. Looking only at the raw uptake rate of a technology can often disguise the rate of change of other associated factors, such as the rule of use, content forms and genres, which can be much slower to develop. The deeper implications of uptake of a technology may only become apparent a considerable time after mass adoption, as business and cultural practices go though profound changes. Alternatively, ideas and images associated with a technology can actually be taken up much more quickly than the technology itself, stimulating other innovations or changing expectations. We thus remain rather uncertain about the overall pace of change – though we can be fairly sure that the predictions by technology promoters of a rapid, across the board uptake of ICT will not be fulfilled.

APPROPRIATION AND THE NORMALISATION OF TECHNOLOGY

The normalisation of technology

Perhaps with hindsight it will be evident that we are currently in a transition period, from an era when ICTs were largely the preserve of a technical elite built around its sub-culture, skills and perspectives, to an era that may qualify better for the description Information society through the uptake and domestication of these technical capabilities by a wide range of groups across society. Norman (1988) has highlighted the possibility of ICTs becoming 'normalised' in social life. A number of ICT artefacts have become largely 'transparent' in their use and utility – for example the telephone, the fax, the mobile phone – and widely adopted. However, as we see below, the key technologies of the Internet are still far from straightforward to install or operate for many users. Norman's critique of the failures of technology design to create computer-based technologies that are indeed readily usable, useful and convivial would seem still to apply (Norman 1998).

ICT applications have not yet become stabilised, 'black-boxed' and transparent in their application and use, in part because of the technological dynamism and turbulence of the ICT sector. Despite 'plug-in and play' claims, there are many technical difficulties in setting up such complex configurations. And learning to use products and services continues to be difficult. There is certainly enormous scope for technologies to be better designed and more transparent, as evinced by the problem that we have described as the 'baroque' design of many contemporary artefacts (such as the video cassette recorder with programming facilities that most users are unable to exploit). However, our study highlights the fact that issues surrounding the development and use of these complex ICT applications cannot be reduced to a matter of design, but depends equally upon wider social learning processes about the potential utility and use of supplier offerings, in which certain capabilities, presumptions and routines for using artefacts become widely communicated. Many computer interface designers quite sensibly suggested that keyboards on computers and mobile phones would severely limit their diffusion, forgetting the ability of people to adapt to the technical world with enough encouragement or coercion. We now turn to examine such metaphors and genres and how they may emerge as a resource for design and appropriation efforts alike.

The social learning of genres and metaphors

In our discussion of technology design/representation, we drew attention to the fact that design never arises from scratch but takes place and is conditioned within a context of the prior history of technology design and domestication. As Carolyn Marvin's study of the uptake of nineteenth-century technologies beautifully shows, the discourses surrounding the design, promotion and use of new technologies take place on a terrain constituted by incumbent technologies. New concepts and practices may come to confront and transform the old in a complex interaction in which 'new practices do not so much flow directly from technologies that inspire them as they are improvised out of old practices that no longer work in new settings' (Marvin, 1988: 5).

In Chapter 5 we also noted the unevenness of progress in developing common understandings about navigation devices, tools and overall narrative structures and genres, that could help to inform the user's navigation around, and use of, the system as well as designer efforts. In particular we pointed to the rapid uptake of certain design/use metaphors conveying the operation of navigation devices (buttons, windows, scroll bars) particularly those associated with the World Wide Web.

It is instructive to consider how this global familiarisation has been achieved. Partly these have become disseminated as common components of computing – e.g. the various elements of graphic user interface. We also see important processes of reinforcement through the transfer of devices/concepts of use between different kinds of application (e.g. the idea of windows for searching transferring from searches within a computer and searches on the World Wide Web). Another, and perhaps underestimated, vehicle has been the transfer of elements between media. Television, in particular, has been quick to take up and popularise certain elements. For example, TV programmers (especially those programmes geared towards children and the 'youth' market) were quick to include windows and buttons in the screen presentations (paradoxically in the case of on-screen buttons which the viewer cannot yet utilise with broadcast TV). Cross-media reinforcement – making TV seem more modern and ICT seem more culturally attractive – may play an important role in the borrowing and evolution of genres of ICT use. They also reflect the fact that broadcast media producers were one of the earliest users of advanced ICT technologies – and quickly learned to incorporate these into their programmes.

Buttons and windows provide rather discrete and simple metaphors for how to use a computer-based system – and ones that can be readily understood and applied in a whole set of areas. When it comes to developing richer concepts of how people might find a system useful, more complex metaphors may be needed. However, one might expect that, under these circumstances, the processes whereby they become generalised may be more obdurate, complex and uneven. Some of the most effective early interface design metaphors are 'physical metaphors' rooted in preceding physical artefacts – for example buttons derive from the mechanical operation of a device. As the information infrastructure becomes ever more complex, the navigation challenges shift to issues of the organisation of information and to visualisation of relationships between people that are more abstract and have few physical correlates. Widely adopted synthetic metaphors may prove valuable as navigational tools (e.g. the increasingly ubiquitous 'search' box). Indeed, new metaphors from the electronic world may even be taken back into physical space.

An effective design metaphor may provide a means of conveying more widely ideas about the meaning of a novel artefact – about its utility and its use. Thus the idea of the digital city has proved a readily understandable device for communicating the possibilities of the Internet as a starting point for organising a virtual community. Indeed, it proved to be an attractive device that could not only inform the activities of technical enthusiasts and wider publics, but also policymakers and the media. The digital city metaphor was particularly important to the Amsterdam Digital City (DDS), helping the project move beyond its founding constituency, a small sub-culture of technology enthusiasts, and making it understandable to lay-people; something that journalists could write about. The success of DDS cannot be divorced from the way it became a cultural phenomenon (van Lieshout cited in Lobet-Maris and van Bastelaer 1999). In this way we can see well-deployed metaphors as effective means to promote the faster, smoother and perhaps more concerted appropriation of novel artefacts.

Metaphors can, of course, obscure and confuse as well as make transparent. They may, for example, act as blinkers and displace or impede the articulation of alternative metaphors and paradigms. Researchers from the Human Computer Interface domain have warned against the adoption of metaphors in artefactual design (Norman 1988, Hutchison 1997, Benyon and Imaz 1999), pointing to the 'metaphor trap', where the application of a powerful metaphor in a design necessitates others features to be aligned with this image even where it is inappropriate, as well as 'broken metaphors' where the evolution of the functionality or usages of an artefact may result in divergence from a once appropriate metaphor (Hutchison 1997). This caution seems to arise from their attempt to see in an interface metaphor some kind of guarantee that ambiguities in the perception of the operation and use of the computer interface can thus can be overcome. Though such arguments may have a certain purchase in relation to reliability engineering (especially in relation to interface design and navigation tools), from our perspective they adopt a rather unhelpful and narrow view of the ways in which metaphors operate as an individual cognitive process. Moreover, whilst reliability engineering may pursue the stabilisation and alignment of understandings of computer use, a social learning approach may suggest different approaches based upon the communication of design solutions that proved attractive and effective to users and in this way develop a broader culture around technology development and use.

A social learning perspective thus suggests a looser view of metaphor. We discussed in Chapter 5 the way in which metaphors could unite designer and user understandings. But this is not achieved by a mechanistic alignment of understandings which it would not be helpful (even if possible) to pursue. Metaphor and analogy cannot be eliminated, since all design representations are inevitably metaphorical – in so far as they operate by analogy and extension. Designers and users are very creative in making these metaphorical leaps. Metaphors are fluid. An important feature of these metaphors seems to be their applicability across a range of situations – on the one hand, because they are generic or readily generalisable and, on the other, because of their scope for flexibility in interpretation – irrespective of the diversity of contexts and perspectives. Their ability to work in this way seems to be underpinned by richness and heterogeneity of these metaphors – they articulate generic elements which many different actors can 'buy-in' to, offering a multiplicity of 'hooks' onto which local actors can project meaning. In this sense metaphors do not resolve uncertainty, but provide a meeting place for different communities. In this way we can see them as 'boundary objects' in that they are 'both plastic enough to adapt to local needs and constraints of the several parties employing them, yet robust enough to maintain a common identity across sites'. (Star and Griesemer 1989: 393) The metaphors deployed in system design acquire a life of their own – no longer rooted in the physical analogy that they were derived from. There are certainly risks that local learning will lead to the breakdown of largely shared understandings or of their fragmentation (especially if local differentiation prevails over dissemination). Van Lieshout (1999) discussing Amsterdam Digital City, notes that metaphors (like artefacts) can function as a configuring device prescribing certain uses and prohibiting others. The metaphor conveys the inner logic of the device and serves as a replacement for it (in so far as the metaphor 'stands for' something else). There is thus a risk of misalignment between the metaphor and the phenomenon it stands for. As van Lieshout notes, 'this replacement may be problematic when the meaning of the one shifts without change in the meaning of the other' (van Lieshout 1999: 64).

These processes have crucial implications for the broader and longer-term questions surrounding the transparency of technology. Hitherto, ICT applications have remained largely in the hands of technical specialists, and an enthusiast culture of early adopters. However, this can be expected to change profoundly as ever larger parts of the population 'buy-in' to understandings of the use and utility of ICT products and services (with the establishment and wide acceptance of new metaphors and genres) and acquire the capabilities needed not only to use them but also to deploy and adapt them for their particular purposes. We should bear in mind that these capabilities are becoming much more widely available through various avenues. For example, we note the efforts of educational establishments to utilise ICT capabilities, not only as a technical subject in education and post-school training but also as part of the generic skills offered at school and especially in higher education. Although, as our education studies revealed, most current teachers still lack these skills (and in the short term had little prospects of acquiring them as the training costs would be prohibitive), as new generations of students come through the education system and come to constitute the new generation of primary and secondary teachers, we can anticipate the much wider provision of such skills, to the extent that they become as much taken for granted as literacy and numeracy. Other routes of skill acquisition and circulation are no less important – particularly since many of the skills required to implement and use ICT have an important experience-based element (e.g. regarding how to configure systems together, how to iron out bugs and how to find short-cuts in using packaged systems). Stewart (2001) has pointed to an important informal knowledge economy, as individuals who have attained the experience and skills needed help neophytes overcome the 'barriers to entry' of getting involved with ICT use and development. Similarly, people may apply in their everyday life ICT-related skills and familiarity acquired at work (Sørensen and Spilker 1999, Stewart 2001). Stewart's work highlights the importance of these informal networks in the private and community use of computers – where adoption and use is often supported an informal intermediary; a local expert (perhaps a friend, neighbour, workmate or relative – but someone informally involved rather than helping as part of their job or as a paid service). It is through these generalised exchanges that ICT applications can in time be expected to become part of our general culture – can be fully domesticated in this sense of becoming transparent and taken for granted (in the same way perhaps that the way of using the telephone, and the skills involved, have become taken –for granted and indeed invisible to us today).

Our research found evidence of the dissemination of familiarity with ICT applications and their usage and a growing level of technical skills and competence amongst different groups across society – as illustrated, for example, by the emergence of sub-cultures of computer enthusiast young women (as evinced by the shift *From Spice Girls to Cyber-Girls*: see Case Summary 6.3). This kind of development could have profound implications for the appropriation of ICT and also for broader policy questions and concerns. For example, this has immediate implications for efforts to overcome not only gender inequalities in access to ICT, but also social inclusion/ exclusion more generally. Though acknowledging the importance of existing barriers which may tend to exclude marginal groups, our research draws attention to other dynamic processes that may serve to undermine or transform these barriers rather than simply reinforcing them. Policies addressing social exclusion should be aware of and seek to exploit these potentialities – and should in particular encompass informal social and cultural processes as well as the established policy agendas which tend to be couched in terms of formal initiatives, e.g. for training. Indeed, one implication of the '*Cyber-Girls*' study and the 'turn to entertainment' in ICT (Sørensen and Spilker 1999), is that we should look at the potential use of these technologies for play and other pleasurable activities, rather than just their functional utilitarian applications, to encourage appropriation of these technologies and the acquisition and sharing of the formal skills and the important informal culture needed for setting up and using ICT products and services.

Case Summary 6.3
From Spice Girls to Cyber-Girls: reconfiguring the gendering of computers

This Norwegian study focuses on a group of computer uses that has not received much attention hitherto: computer-fascinated girls who are enthusiastic in exploring uses of the computer and Internet. It points to a relatively successful reconfiguration of the gendering of computers. A key factor was the development of the Internet that has in a very profound manner changed the meaning of computers and made them far more attractive to girls. The computer is used for intimate social communicative roles, such as writing a diary and writing poems. The girls perceive the computer as a communicative device, by itself and as a gateway to the Internet. Surfing for fun, searching for attractive boys and chatting are amongst their favourite uses.

The emergence of a sub-culture of computer-enthusiast girls was underpinned by a change in the gender distribution of computer skills (this is not to suggest that the construction of girlhood has changed, since these young women emphasise their gender normality). These girls acquired their computer skills mainly through learning by doing. This may present a paradox to education policy, since computers in education initiatives often seem to have had the effect of driving women away. To get girls interested in and fascinated by computers we may need to exploit the attraction of playful activities – to explore the potential of the computer as a toy – and one which you can continue to play with as you get older (Nordli 2001).

Stabilisation and the normalisation of artefacts?

Given that most of the ICT applications being developed and used are based upon relatively standard technology platforms (typically a personal computer with Internet access), the level of technical reconfiguration during the appropriation process – and thus, for example, the potential for innofusion in the *technology platform* – could be expected to be rather modest. This does not mean that their application was seamless and without effort. Indeed, these technologies are still difficult to implement and use. They are by no means as 'easy to use' in the way that suppliers would have us believe – that they can simply be plugged in, switched on and used. For example, even a group such as UK Bank's Technology Research Department, possessing high-levels of technical skills, found themselves stretched in their attempts to configure together the notionally standard electronic mail systems, network and operating systems, telecommunications links, and personal computers that they needed for their home-based teleworking trial. Initiatives that promise to deliver hardware and software that are easier to use have been criticised for having little visible effect.[5] At the same time, the continued technical and commercial dynamism and turbulence surrounding the development of technical components can undermine attempts to stabilise technological development. As a result these offerings are often far from finished – particularly in relation to their interoperability. This was particularly an issue around the Internet where competition amongst a multitude of suppliers of component technologies and platforms results in shortened product life cycles and competing and changing standards.[6]

At present the competencies needed to deal with these kinds of technical problems are not yet widely disseminated. Many of the skills required are highly contingent and relate to particular problems in operating and configuring together specific technology components. They involve rather informal kinds of knowledge, that can only be acquired through practical experience. 'Getting started' with a new system typically involves a rather substantial investment of time and effort in configuring a system and sorting out the many 'teething troubles' that arise. Contact with someone who had already been through this process could be extremely helpful. Novices often drew upon this kind of experience through informal channels.[7] Reducing these entry costs was one of the motivations for people to go to cybercafes, where someone else had gone through this 'learning curve'. More generally overcoming such barriers in getting started and learning how to use a system constitutes one of the main reasons for the importance of 'appropriation intermediaries'.

Case Summary 6.4
Coffee and Computers? Cybercafes – a site for the local appropriation of ICTs

Cybercafes provide a clear example of the way that local intermediaries appropriate and reinvent the concept of the cybercafe and the Internet in configurations that respond to the local market, political and cultural agenda, and to technical change and the emergence of new supplier offerings. Three cafes were studied in Scotland, and three in Ireland, include a cyberpub. This was at a time in the late 1990s when cybercafes were regarded in many quarters as dead-ends on the information superhighway (a vision in which new ICTs would be privately owned and consumed)! The cybercafe owners and staff created new spaces and found a market for their services in a very interactive way: they were dealing with a technical and physical place that was relatively unknown, and, although they started with a basic concept, this metaphor had to evolve in order to bring in and satisfy customers. They played the roles of facilitator – giving access to the Internet and computers in a public place, and providing basic training and advice, and the role of configurer, deciding not only which technologies to offer, but also about the décor of the café, and what sort of atmosphere to encourage. In some cases the managers played a brokering role, negotiating with supporters over the purpose and configuration of the café in order to gain funding.

Cybercafes also illustrate processes of learning, both around technology use, and the concept of public Internet access. By creating welcoming spaces associated with leisure, rather than school and work, cybercafes present the human face of computing, associating it with relaxation, play, communication and sociality in a familiar context. As opposed to more 'worthy' public access centres, linking ICTs with gaining skills for work or learning, cybercafes survived and prospered as commercial entities by allowing any uses, and finding ways to satisfy different user groups, from web-surfing grannies to boys playing networked games (Stewart 1999, Laegran and Stewart 2003).

APPROPRIATION AS AN INTERMEDIATION PROCESS

In Chapter 4 we flagged the importance of 'appropriation intermediaries' in mediating consumption and usage and suggested that they may play two key roles in the appropriation of ICT:

1. First is their potential enabling role in consumption and use – in reducing the barriers to access, entry costs and uncertainties for 'users, particularly in relation to non-specialist users.
2. The second possible role of appropriation intermediaries is their potential contribution to innofusion. Every implementation involves some local reconfiguration of systems. This may involve quite a substantial technical effort given the lack of stable standards and methods of operation. Knowledge of these activities can be an important resource for technology developers not only in identifying 'bugs' and problems but also in refining their products and developing new products. Their potential contribution can be evinced by the efforts of suppliers of ICT products to develop more sustained communication channels with their consumers through 'alpha' and 'beta' testing. Appropriation intermediaries could be used as a similar resource.

We shall explore these in turn.

Active and organised appropriation

Thus far we have mainly been considering the informal social learning processes that arise spontaneously around the implementation and use of new ICTs. However, appropriation may need to be an active and organised process, particularly in relation to new kinds of ICT products and services. One of the most interesting examples of the role of appropriation intermediaries in consumption and usage is provided by the 'Virtual Museum of Saint Colm Cille' (see Case Summary 6.1). The developers, Derry Nerve Centre, recognised that uptake might otherwise be slow, limited and uneven, and, before the development of this product had been completed, they embarked upon an active process to support diffusion and appropriation in an attempt to build a market – for example, introducing local schools and community groups to the new and unfamiliar medium. The Nerve Centre case highlights the fact that considerable resources may be needed to support this kind of broader dissemination/appropriation effort. As a small group with limited resources, they sought to utilise existing popular cultural outlets such as film festivals in order to find cheap, low-risk and trusted avenues for disseminating and popularising their products (Preston 1999). One obstacle Nerve Centre encountered was that public support for commercialisation (i.e. for the resources needed to foster widespread appropriation) proved more difficult to obtain than for the initial research and development of new technology products (Kerr 2002).

Craigmillar Community Information Service faced a similar challenge of actively enrolling new user groups. Though its starting point was established community groups, these showed little interest in using the system as initially conceived. CCIS staff changed their strategy and sought to foster the development of particular groups of user such as the senior Citizens (the CyberGrannies). What is of interest is the way that the potential relevance of CCIS was constructed around each group – as we noted earlier, the senior citizens were assumed to want to use the Web for 'remembering' and to contact groups of their peers. CCIS also linked its information system up to external (e.g. North American) community information systems to make the system more relevant (a move which promoted controversy about the proper users and uses of the system [Slack 2000b, Slack and Williams 2000]).

The final cases we shall discuss here involves the cybercafes (Stewart 1999). These were public sites for appropriation of new commercial ICT offerings for entertainment and everyday life. The cybercafe managers were very active in creating an appropriately designed and attractive setting – a safe and enticing place where particular groups could engage with new supplier offerings (Laegran and Stewart 2003). They created an ordered context for the appropriation of novel ICT offerings.[8] The managers were also active as market mediators – searching out what new games and other products were becoming available, evaluating them and selecting appropriate offerings. Finally they played a role in implementation and final appropriation – installing products, dealing with technical problems, and learning how to use them – thus reducing barriers to entry/foreshortening the learning curve for new users to get involved with these various offerings (as well as, post-implementation, helping to build up and exchange local knowledge about how to play). Cybercafe managers thus provided a very important bridge in the appropriation process. They constitute particularly salient instances of the 'appropriation intermediaries' that we argue are critical in the current early stages in introducing new ICT products. In this case their role was commercially motivated – users were willing to pay them (through usage charges) for their role in reducing search/access costs, reducing risks and sharing costs of acquiring new commercial products and, in providing an accessible and convivial social setting, for adding value and meaning to commercially available products. The cybercafes in our studies did not seem, however, to have been engaged as a vehicle for direct feedback to future technology supply. This kind of context for collective and semi-public consumption might be an important resource for suppliers to tap in order to secure appropriation experience.

Unanticipated use – and its potential as a resource for innofusion and domestication efforts

Our studies highlight the complexity and unpredictability of the appropriation process. In virtually all cases where we were able to examine implementation and use of ICT products and services we found more or less clear discrepancies between the concepts underpinning the design of the product and the richness and diversity of user appropriation experiences. Unanticipated uses and outcomes were frequent. And whilst new technologies are often conceived around systems to support existing activities, one of the clear retrospective lessons from technology studies is that the main benefits that accrue from a technology are often far removed from these preconceptions and involve the more radical changes in usage and behaviour that emerge through domestication and social learning processes. We will review some of these findings here. They confirm our insistence on the importance, and active nature, of the appropriation process. From the point of view of social learning, a key question concerns whether and how we can integrate knowledge emerging from appropriation processes into future technology supply.

Examples from educational technology

As already noted, the cases in which we were best able to explore in detail appropriation as well as technology design were mainly ICT applications in organisations and in education, where there was a well-established setting linking these contexts. We will start by exploring our educational ICT experiments, where we have fairly systematic information about appropriation.

For example, even though the Language Course case involved users in a tokenistic manner (teachers and students were brought in so that the pilot could be demonstrated at a forthcoming trade exhibition) it is striking that, despite this, the Language Course developers DID learn important lessons from these users about how their systems might be used in different ways to those anticipated, and the consequent need to reconfigure the system. For example, the designers had been working with the presumption that the children would only use the system in the ways they anticipated – overlooking the rather well-known propensity of children to play and their creativity in inventing playful uses of the system. In the original design, a voice switch was used to control the source of the video-channel broadcast. The children invented a game of making noises to bring the attention of the video-conferencing systems to them and then hiding. To try to regulate the behaviour of the children, the teacher was given a device to turn off the child's microphone and the children were given a button to alert the teacher if they wanted to speak (Mourik, 2001).[9]

Often the unanticipated uses that emerged related to working around the shortcomings as well as the affordances of the system as installed. Again with Language School the time delay between the receipt of the audio and video signals (which arose because of the limited telecommunications carrying capacity) ruled out its intended use for pronunciation exercises. The teachers had to find new ways of teaching around the affordances of the system. The medium – for example of the electronic whiteboard and voice – was available to all. Teachers could not teach individually because then the others would not be able to learn, so the only way to help the slow pupils was by giving them more advice and less difficult assignments. Education changed from being teacher centred to student centred. The teacher became more of a facilitator and mediator between the information available in surplus through the system and the student. Each student, in other words, defined their own course and rate of progress (Mourik, 2001). The Danish Bornholm tele-education case identified similar types of change in the role of the teacher. Whereas some had expected the new medium to 'deliver' education by the teacher, the latter needed to adapt teaching methods and materials to this new communication medium (Hansen 1998).

Case Summary 6.5
Teleteaching in Bornholm – learning new means of interacting for new media

The Bornholm Teleteaching project shows how users have to develop new means of interaction and presentation when moving to new media. The project was designed to deliver teaching to a relatively isolated island off the Baltic coast of Denmark through the use of video telephony. Social learning dimensions centre around the need to translate the teaching content to the technology and to develop pedagogical styles that suit the medium. Practically, the introduction of a distributed teaching and learning context profoundly changes the relationship of teachers and students.

Students and teachers are not co-located and there is a need for greater awareness on the part of teachers about classroom management issues:

teachers have to 'host' programmes on the video-conferencing system as opposed to teaching as if the students were in the same room. The usual rules for conversational interaction do not apply in such environments and *new formal rules need to be developed* if the lesson is to proceed effectively. Such simple tasks as asking questions are made problematic in that the technology does not facilitate the use of gaze and intonation to gain students' attention and to indicate that the question is for them. Teaching materials must also be changed to facilitate the use of distance teaching and learning in that the materials must explain fully what is required since the teacher will not always be available to elaborate on what the task demands. Teleteaching in Bornholm led to a change in the roles of teachers and students. The project shows that it is no simple matter to translate content across media since this involves a switch of genres and attendant representational conventions. To be effective, new skills must be learned by all involved (Hansen and Clausen 1998).

Broadly similar sets of issues came up in the Telepoly case using a high-speed broadband network for distance teaching in the Swiss higher education sector. Some teachers failed to adapt their teaching methods to the new medium, and consequently received negative feedback from the students and lost enthusiasm for the project. Two lecturers were successful – though in very different ways. One adapted to the particular style of communication imposed by the system, such as addressing the mobile camera, talking slowly, and thinking constantly about possible interactions with the virtual classroom, visible only on a TV screen. He had put a lot of effort into preparing this teaching including putting documents in electronic form for the whole course. The other lecturer installed his teaching on Internet with many hypertext links with demonstrations, which the students could consult whenever they wanted (Buser and Rossel 2001).

In the Cable School experiment in the UK (Case Summary 4.3), the eventual outcomes were far removed from the initial conception – teachers in different schools would use the system for sharing worksheets and other resources electronically. This was defeated by the failure of the initiative to provide resources or institutional support and incentives for teachers to develop such sheets.[10] However, at the end of the project, some uses of the local servers and networks had emerged become institutionalised within the participating schools, including email and the collection of World Wide Web pages as a teaching resource (Slack 2000a).

What is clear is that, although considerable progress has been made, we are still at the early stages of the appropriation of ICT within education (and elsewhere). The key challenges still surround the articulation of technical and pedagogical potentialities – and there have been some pointers from our cases in terms of the shift to student- and resource-centred learning approaches. Many of the educational initiatives we have examined have been primarily in the hands of technology specialists or enthusiasts – whose technically driven visions have often proved 'out of kilter' with the circumstances and culture of most players in the appropriation domain.

It remains the case, as noted earlier, that a key prerequisite for more effective utilisation of ICT in education will be the involvement of teachers (and through them the students), in exploring and developing the potential of these media in different local educational settings. The potential contribution of teachers as actors in this process (rather than simple recipients of technological offerings) has not been properly recognised. Indeed, van Lieshout, Egyedi and Bijker (2001) describe the support that teachers receive as 'shameful' – noting that limited funding does not allow for training and support of the average teacher, and that enthusiasts from teaching staff often contribute in their in leisure time. Innofusion and other appropriation lessons could easily be lost because teachers were not properly institutionalised within project management structures that focused upon commercial suppliers and education managers. There were, however, differences in this respect between our cases. There was, for example, some indication that teachers in higher education tended to have more autonomy, and better access to skills and other resources needed to shape the application of ICT, than for example those in school settings.

Insights from other cases

These kinds of creative efforts of innofusion and domestication were by no means restricted to these educational technology applications, but was found wherever ICT was being appropriated. This underpins our argument that ICT artefacts remain unfinished – in the sense that they have not become stabilised or fixed in their meanings and usages. Users made different interpretations of the meaning and uses of artefacts depending upon their context and standpoint. In some cases they elaborated uses exploiting functionalities of the artefact that were not foreseen (or even implied) by the designers, or changed the understanding of existing functionalities and their use. For example, the UK Bank's video-conferencing experiment showed them which of the bundle of functionalities they had configured together from commercial packaged solutions would prove helpful for remote cooperative working. The case also involved 'social innovation' through the elaboration of new sets of social conventions for such video-mediated communication – mainly to find ways of working around the restrictions imposed by the technology (e.g. difficulties of distinguishing people in group meetings; the need to take turns in speaking and regulate who would go next; the need to prepare in advance for such meetings) (Procter, Williams and Cashin 1999).

Case Summary 6.6
UK Bank – experimenting with desk-top video-conferencing (DTVC)
A series of pilots of DTVC conducted by UK Bank constituted a test-bed and an arena for experimentation in the configuration and use of ICT products.

The study highlights the choices available to the users in the configuration together of a range of standard packages and tools and their creative role in appropriating the new functionalities into working practices. 'Users' from the Bank's Technology Research Department and the business divisions involved took decisions about which functionality was relevant to them – only some of the options provided for in the standard packages and tools were used and others were simply switched off.

In learning how to work with the new functionalities offered by DTVC products, users explored which of the embedded options were suitable for particular kinds of activity. For example, though the video-link was important in establishing a sense of closeness with a remote site, it proved less important in working collaboratively than file exchange, and the ability to simultaneously view spreadsheets. Indeed, when working in this way, users would often switch the video icon off, as this interfered with their attempts to discuss spreadsheets remotely, leaving the screen space free for the data. Users also had to learn new sets of social conventions for such video-mediated communication – in particular to find ways of working around the restrictions imposed by the technology.[11] DTVC is still used mainly for formal meetings, which the participants have booked. People have to prepare for the meeting in advance, in particular to decide what data they needed to transmit. Understanding of the uses and significance of these technologies changed. For example, in these pilots, DTVC moved from the manager's desk to a public access area.

Similar kinds of experimentation were also encountered in some of the cultural product developments we studied. In particular The Den on the Net (Case Summary 6.7) shows how RTE used websites and a number of high-profile demonstrations to learn about new technologies and new digital media. The Web also provided a way of fostering links with its audiences – including the overseas Irish diaspora.

Case Summary 6.7 – RTE: The Den on the Net: global technology traditional media and cultural content
This case study focuses on how RTE. a traditional national public service broadcaster of TV and radio, is experimenting with on-line services in an attempt to expand upon its existing media services and potentially develop new revenue generating activities. This study shows how the development of websites and a number of high-profile Internet-based experiments and one-off demonstrations (e.g. broadcasting of music concerts) provided a space for RTE to learn about new technologies, new types of content and new ways of interacting with the audience/users.

The case study highlights the technical and economic limitations and opportunities for creating digital content faced by what is (in global terms) a relatively small media company operating within a minority cultural milieu, in an era of increasing global control of cultural goods. The initiative was rather successful in establishing RTE as a key national player in the new digital media field (measured for example through the number of hits/users at national/domestic and international levels). In addition the websites provide a means of communicating with (as opposed to just broadcasting to) the Irish abroad, the 'diaspora' market. It also indicates how the new digital medium is used to foster an on-line community of interest comprising Irish nationals, Irish emigrants and people from other nations interested in Irish 'culture' (Kerr 2000, 2002).

The innofusion processes we noted went beyond what has normally been flagged in discussions of 'supplier–user links' (where knowledge about user responses to supplier offerings, flowing to suppliers, supports product enhancement – overcoming the limitations of feedback through the impersonal market). The flexible and malleable character of artefacts that are based on ICT, means that the traditional borderline between innovation and diffusion has become blurred. This was particularly notable in cases such as Amsterdam Digital City (DDS), which can be taken to exemplify an evolutionary model of system development, which went through a process of continual development and one which involved many players – including media people and everyday users as well as technical specialists – through a shared culture (van Lieshout, cited in Lobet-Maris and van Bastelaer 2000).

This kind of feedback from the contexts of use to the context technology supply that may result from innofusion often – indeed mainly – takes place through *informal channels*. Many players are actors and contribute to this process. This kind of learning is hard to plan for and predict. What is learnt is often unexpected, serendipitous and highly contingent to particular settings. Questions then arise about what avenues exist to carry and deploy this knowledge more broadly: as a resource both for the further appropriation of the artefact and for future technological supply. The strength and effectiveness of these channels varied from case to case. There were also marked differences in terms of how *widely* the knowledge and experience was communicated and the *persistence* of such knowledge.

ICT trials and digital experiments constituted relatively organised and well-structured/supported spaces for these different actors to meet and communicate. Intermediaries – who often emerged spontaneously in those cases there was no formal support for this role – provided an important avenue for the further exploitation of these experiences in the form of embodied knowledge. There were, however, particular issues concerning feedback from final-users, who may be far removed from settings of technology development spatially, temporally and in terms of their societal roles. There are lessons about the organisation of digital experiments that we discuss in the next chapter.

In this section we have focused mainly upon the substance of what was learnt in social experiments or commercial trials, such digital experiments *are about* ***process*** *as much as substantive outcome*. Social learning has multiple dimensions. Learning relates both to representations, designs and uses of new ICT products/services, and to the processes whereby these are developed and appropriated. Indeed, learning about 'the technology' turned out often to be only a relatively minor component of the experiment; finding out about the behaviour, concerns and commitments of other players (suppliers, intermediate and final consumers, promoters, regulators and policymakers) proved to be a more important outcome of many experiments than the narrow technical outcomes. This was the case even though it was often the technical goals that figured most centrally in the formal objectives of the project. A project may bring varying kinds and levels of opportunity for social learning along different dimensions.

Trials and experiments are thus a way of learning **how to** develop new ICT applications as much as they are about **what** these products and services will look like. This involves learning, for example, how to build constituencies of suppliers of component technologies (e.g. communication networks, terminals) and complementary products (particularly in relation to content); how to obtain the necessary resources and build commitments and alliances with players inside and outside the constituency, as well as how to conduct user trials etc.

Many of the most important lessons from an experiment may be more general reflexive ones: about **how to** learn through experiments about constituency building, how to address uses/users etc. This has important implications when we come to consider the implications of this study for public policy/business strategy in Chapter 9. But first let us consider the implications for the organisation and management of digital experiments. The next chapter will explore the issues surrounding the management and organisation of digital experiments themselves and the relationship between the project and its context. As well as addressing the conduct of social learning in the design–implementation–use cycle it also raises the broader processes of learning by regulating around these experiments.

NOTES

1 Social dispersal is particularly an issue in relation to domestication; innofusion processes in contrast are more focused spatially and temporally at the point of implementation
2 Four cases by each of seven SLIM Centres plus Edipresse and UK Bank – though the latter two both contain more than one case each.

3 In the case of Edipresse and UK Bank some applications also involved relatively standardised information services (typically based on the World Wide Web) for a wider array of customers.

4 This, and the subsequent methodological and resource constraints, was one of the reasons why the SLIM project was not able to address the consumption/appropriation of ICT *in the home* to the extent that was anticipated in the original research design. Another factor was that we did not find the expected roll-out of ICT products to widespread domestic use.

5 For example, various industry initiatives, like Easy PC, led by Intel and Microsoft, that claims to deliver 'Software that's easier to use ... (on) hardware that's easier to operate', have been criticised for having had little visible effect (Jack Schofield, 'Windows 98 reprieved', The *Guardian On-line* section Thursday 15 April 1999).

6 Set-up and operational problems are far lower, of course, in the case of mobile phones, despite rapid service and technology changes in this sector, as with the Internet, as these continue to be dedicated devices, operating in a more regulated context.

7 Individuals have different levels of access to such local experts – though they turned out to be a very important – for example in the hypertext programme used in the IMMICS media education course, particularly for those students who were less committed to and experienced in the use of these technologies.

8 In the cybercafes, as in the case of DDS, an important element of this appropriation process concerned the establishment of formal or informal rules for how participants (citizens or customers) should behave. Appropriation in collective settings involves certain types of decision about what kinds of behaviour are proper and appropriate. Such local rule-making draws upon resources from the broader setting of norms and regulations. We point to the reciprocal interaction between wider rule-making activities/regulation and local regulation in creating and sustaining technology regimes.

9 It must be noted that this represents a rather mechanistic response to the 'problems' of unanticipated use, based on technical constraints rather than developing social conventions to regulate computer-mediated interaction.

10 It seems paradoxical, given that Cable School rested upon collaboration between schools, that there seems to have been little attempt to create a supportive culture and norms between the subject teachers involved across various schools. There were a few meetings of teaching management in the various schools involved but no direct contacts between the subject teachers who needed to be at the heart of the experiment. These teachers told us they felt inhibited from presenting their offerings on the system, for fear they might attract critical comment from peers in other schools. This provides yet another instance of the damaging consequences of the rhetorics of technology supply evinced by technology technical focus of the project; the conceptualisation of technology as a solution to social problems; the idea that simply installing technical capabilities would generate change in teaching culture and practice.

11 In its initial implementation, the UK Bank video-conferencing project also encountered problems with voice-switching of the video channel that bears some similarities to the Language Course case. When the system was used in half-duplex mode – grunts of affirmation would unintentionally transfer communication from the speaker to the listener.

7. The Conduct and Management of Digital Experiments

As Mackay et al. (2000) point out, the boundaries of a project and the relationships between designer and user are not fixed but must be negotiated, managed and indeed configured. This chapter seeks to analyse the choices in the way that digital experiments were conducted and managed. We take as our starting point the broad model and mapping of the terrain for social learning already articulated in Chapter 4. Our discussion highlighted, for example, the processes through which ICT constituencies (or constellations) were developed, the active role of intermediaries in bringing together players with diverse interests and perspectives, to establish visions and incentives to motivate their involvement in order to secure the intellectual and material resources needed. From our own and related studies we can identify a more specific set of recurrent issues and dilemmas concerning how the boundaries around digital experiments were set and managed. This involved choices about which players were involved and in what way – with some clear differences across a range of very different ICT experiments.

These choices can be divided into two broad categories:

- The organisation of the design/development process in terms of i) the level of control exercised over the project, ii) the range of players involved, and especially the extent and manner of involvement of 'end-users' and iii) how the relationship between designer and user is conceived
- The relationship between the project and its context – focusing in particular of the efforts by project teams to create a supportive environment and the role of public funding.

THE ORGANISATION OF DESIGN AND DEVELOPMENT

In conducting a digital experiment there are important choices to be made about how the parameters of the initiative should be set in terms of the goals adopted, how open the development should be, and how and when closure should be established. First, across our range of studies we identified two contrasting ideal types of design/development that we characterised as *mode of experimentation* and *mode of control* (van Lieshout 2001, Egyedi and Bijker, van Lieshout 2001). We found our different ICT projects are located along a spectrum between experimentation and control, and unpicked a number of specific dimensions of these two models. Second, we found that one of the key issues concerned participation – who was involved and how – particularly in relation to *involving the end-user*. Despite the wide espousal of the importance of 'involving the user', our cases highlighted the frequent lack of such involvement – certainly in terms of the direct involvement of the user in design/development teams – and the difficulties in such an endeavour. Third, we found differences in the way that design was conceived in terms of the designer – user relationship, between discourses that emphasised the *creativity* of the designer versus the *accountability* of the designer to the user. We will address these in turn.

Mode of experimentation and mode of control in digital experiments

Our research project has investigated a wide range of ICT initiatives mainly located across three broad application domains (Education; Public Administration; Cultural products). An important empirical variable between cases, and between domains, concerned their location on a spectrum between what we have described as the Mode of Experimentation and the Mode of Control (van Lieshout, Egyedi and Bijker 2001, Lobet-Maris and van Bastelaer 1999, Preston 1999). This relates to the level of openness of the project both regarding participation in the project (and the scope for extending participation) and regarding its goals. In the mode of control, typically, there may be an attempt to define the goals and expected outcomes of a project in advance and rather precisely, and also to prescribe who will participate in development decisions and the resources used. In the mode of experimentation, there is provision for a more open learning process, with greater openness openness to other participants and agendas. The benefit of the mode of experimentation is a greater willingness to change direction and approach in the light of experience; the 'down-side' may be (at least perceived) uncertainty and loss of predictability and control over the direction and outcomes of a project.

van Lieshout, Egyedi and Bijker (2001) have distinguished a number of different dimensions of the modes of experimentation or control. Particular experiments may be differently located on these dimensions on a spectrum between experimentation and control. These elements, which may be more or less closely related, are summarised in Figure 7.1.

Figure 7.2 seeks to display these different dimensions of the mode of experimentation and control on a single diagram (capturing for comparative purposes the various educational technology studies [van Lieshout, Egyedi and Bijker 2001]). For each dimension, proximity towards the control end of the spectrum is at the centre of the diagram, while more experimental approaches are shown at the edge.

Mode of experimentation	**Relevant dimension**	**Mode of control**
Open to change	***dominant attitude***	semi-open to change
Process oriented	***goal/objective***	product oriented
Minimum of boundaries and constraints	***environment***	explicit boundaries and constraints
Exploratory experimentation	***mode of practice***	purposeful experimentation
Open style of management	***style of management***	hierarchical style of management

Source: adapted from van Lieshout, Egyedi and Bijker (2001: 291)

Figure 7.1 Dimensions of Mode of Experimentation *vis-à-vis* Mode of Control

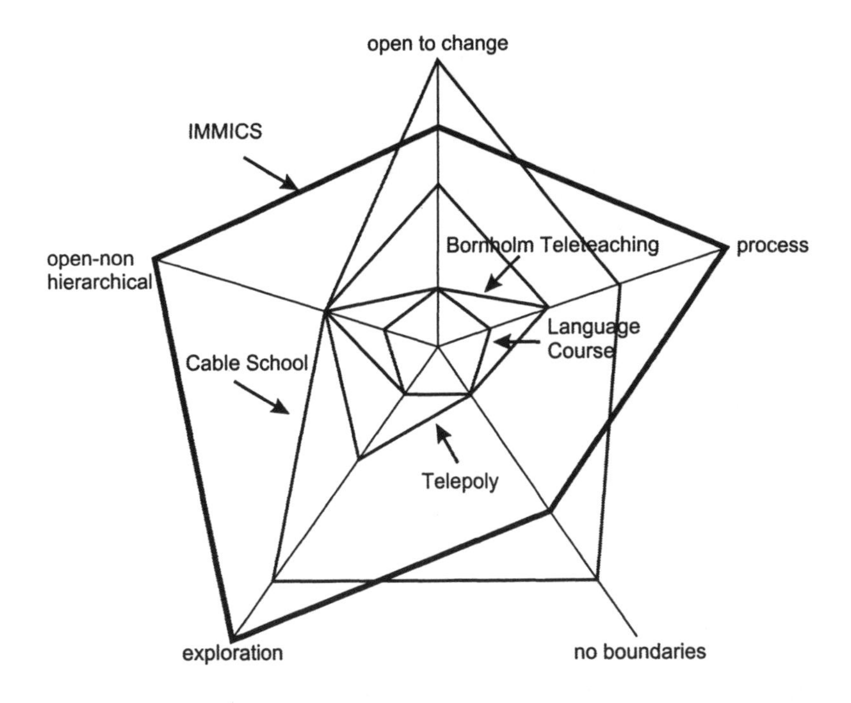

Source: adapted from van Lieshout, Egyedi and Bijker (2001: 291)

Figure 7.2 Diagram of Dimensions of Experimentation and Control

Control is of course very much a question about ***who*** is seeking to exercise control and over what issues. Figure 7.3 seeks to clarify the contrast between digital experiments conducted under the mode of experimentation and the mode of control by showing what these might look like in relation to the space for social learning that we established in Chapter 4. It shows how these differences in control and openness relate to two key dimensions: first, the *range of players* involved and, second, the *extent of their involvement* in and control over the various stages of project development. In other words, in the mode of control (illustrated by the rectangular box) control over the project is narrowly restricted (perhaps to a narrow technical or managerial group), and broader arrays of users are doubly excluded – both by not being party to the experiment and because the key design decisions have been taken prior to the roll-out and use of the system. This exclusion and preclusion is less clear in the mode of experimentation (illustrated by the oval).

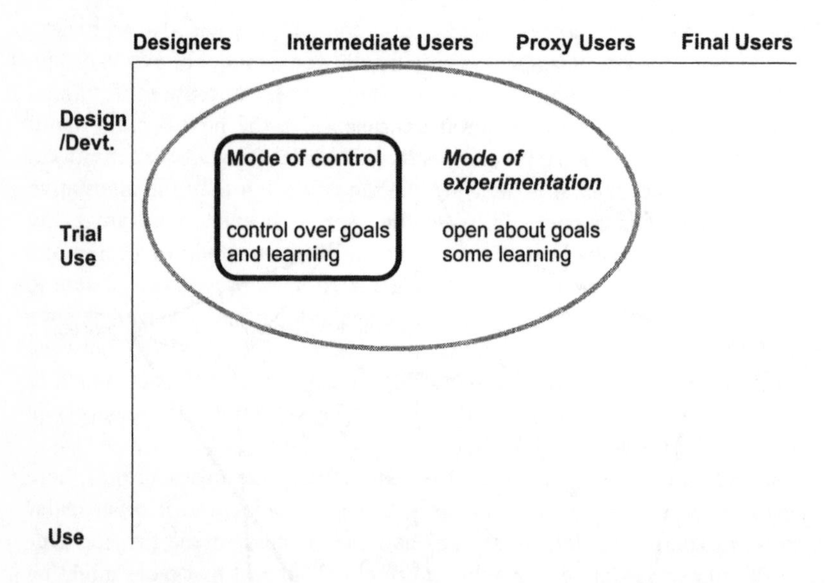

Figure 7.3 Mode of Experimentation and Mode of Control in Digital Experiments

In some of our cases the adoption of the *mode of control* seems to reflect particular contingencies for the digital experiment, notably in the case of projects with a stronger technical development focus, where the mode of control corresponded also to the imposition of a narrow technical agenda and formal division of labour. There are presumptions in such projects about the adequacy of available technologies to meet user needs, encompassing both how well understood user objectives are and how readily these can be met by design. The apparent presumption that both of these can readily be resolved is reflected in what van Lieshout et al. (2001) describe as 'verification experiments' in contrast to 'diversification experiments' conducted under mode of experimentation with greater openness about the technical tools available and the problems being addressed.

One example of mode of control was the Pericles public administration application. The adoption of this mode can be related to one of the goals of the project – to develop an on-line front-end to the local authority's internal information system which involved a considerable development effort. However, this case was somewhat exceptional amongst the public administration and information systems we examined. Most of the other cases studied were not so tightly geared to particular information systems and organisational/administrative practices, but were relatively independent stand-alone systems for information sharing.

The development of some of the cultural content products also seemed to resemble the mode of control in so far as it followed a model of expert design – in so far as cultural production is often carried out by professional designers. This links in to the point we discuss in the next section about different *modes of design*. The more authorial model (mode of creativity) of design may also suggest a greater privileging of design than the alternative mode focused upon accountability to the user. However, there are many dimensions of control – and the relationship between mode of design and mode of control is unlikely to be simple. Indeed, though we may be able to draw some associations between design modes and particular exigencies, we should not overlook the evident scope for choice that existed between experimentation and control. Thus, rather similar types of projects (such as our various digital city/public administration cases) could differ sharply in this respect (Lobet-Maris 1999).

Another immediate empirical observation from the cases is that there appears to be no simple relationship between the selection of a particular mode (of experimentation or design) and the success of the project. Our concern with social learning would, in principle, suggest that there might be important learning advantages from a more experimental approach. On the other hand, we have already noted the potential costs and uncertainties that may arise from the latter's unstable and dynamic development setting.

The calculus of risks/costs and benefits within a project, of course, represents only a part of the overall assessment. Indeed, as Figure 7.3 reminds us, the formal conduct of a digital experiment, and the way its goals are initially set, only forms a part – and perhaps only a small part – of the broader domain for social learning in which IST applications will ultimately be innovated and appropriated. The conduct of experiments thus needs to be addressed against this backdrop. The appropriateness of choices, for example between modes of experimentation and control, can be only partly assessed in the light of the experiences within a project, and may ultimately depend upon the outcomes within this broader domain. On the other hand, as Sørensen and Spilker (1999) point out in relation to mass-market products for the home, designers committed to the mode of control will tend not to be interested in examining social learning when their products are consumed; 'users are seen as a challenge of persuasion, to make people buy the product, not as a resource to make improved designs'. (Sørensen and Spilker 1999: 72)

Much depends, therefore, on subsequent, broader processes of social learning outside the formal boundaries of the experiment. Our research has pointed to the success of the evolutionary development model (exemplified by DDS), in which the experimental approach is maintained through subsequent (re-)design iterations throughout the appropriation as well as on the prior development stage. What may be at issue here may revolve around rather different views of the life cycle of an ICT artefact as, on the one hand, a finished product and, on the other as a product that remains perennially unfinished. Whilst some products may become stabilised and 'transparent', the overall thrust of our studies is that many of the ICT applications and usages being developed across such diverse settings as education, community information and cultural content remain profoundly experimental.

Case Summary 7.1
Hypermedia Use in Education: a diversification experiment

This case is about the use of a hypermedia programme for a new visual culture course. The course has two aims. Students are to learn how war is represented in different visual media (e.g. photography and television), and they are to learn how to do research in a non-linear way. The use of hypermedia software is a means to make explicit the process of 'non-linear' structuring of research data. The software is new to all concerned.

Two teachers treat hypermedia use as an experiment. They do not define a best practice beforehand. They seek diversified hypermedia use and therefore steer as little as possible. Their attempt to exploit the interpretative flexibility of hypermedia use and postpone 'closure' is a main ingredient of what we term *diversification experiment*.

It proves difficult for teachers not to give any pointers at all for the use of hypermedia for non-linear research. They partly succeed. At times, the lack of guidance frustrates students. The students' capacities of home computer user and experienced learner affect their hypermedia use in the course. Some hypermedia uses are constrained by the software. In certain cases students find ways to work around these constraints. Unexpected hypermedia uses develop. Postponement of closure appears to encourage new uses. Furthermore, it requires frequent reflection on the use of hypermedia for non-linear research. The experiment has led to intense learning by students as well as teachers. Highlighting the divergences between stated purpose and end result, the case illustrates how Fleck's concept of innofusion can contribute to our understanding of the processes of negotiation that surround such developments.

Our emphasis on experimentation should not obscure the importance of some degree of control and coordination in ICT projects. Indeed, the lack of a central player able to coordinate or control these experimental efforts could also prove problematic, as exemplified by the Cable School (Slack 2000a) and other cases (Nicoll 2000). The overall lesson from these discussions is that what is required is an appropriate *combination* of experimentation and control. As our study of educational technology concluded:

> Control is needed to survey and evaluate what is going on, to offer clarity of responsibility, and to motivate those involved by showing innovative uses and by offering a platform to exchange ideas and experiences. Exploration is needed since educational technology are no *'one size fits all'* products. They have to be explored on their possible uses, they have to find a place in the setting of use, they permit new and innovative uses that were not foreseen. (Bijker et al. 1999 ch. 13)

Both experimentation and control are thus important. However, there is a sense that technically focused projects have often veered unduly towards the mode of control, and that the necessary experimentation has had to be 'smuggled in'. Technological objectives have taken first place, and only after they have been realised has space emerged for experimentation about usages (for example in the above cases, about educational objectives [*ibid.*]). The presumption that the technology would provide a solution per se, meant that users have had to grapple with the constraints and affordances of new technologies under circumstances of use and according to the parameters configured in design (for example in teleteaching with the Telepoly initiative). There was a failure to provide sufficient time and a safe context for users to 'play around' with the technology. In the case of Telepoly (and likewise Cable School) the result was that teachers who had negative experiences lost their commitment to the project. Thus one of the important strategic choices in building and conducting a digital experiment concerns the dimensions and timing of control.

The involvement of final-users

Another key question surrounding the 'configuration' of a digital experiment concerns which actors are involved, in which way and at what stages of development, and thus how they may contribute to design/development choices. Here considerable attention has been given to the involvement of 'final-users' in system development.

The public rhetorics of ICT have been emphasising the centrality of the user for over a decade – and the idea of user-involvement has become almost a *sine qua non* in the contemporary teaching of computer systems development. How do we reconcile this with one of the most striking and surprising findings noted in Chapter 4, concerning the **marked absence of constellations directly bringing together end-users as well as supply-side players** across our wide range of case –studies?

This absence raises a raft of questions that we are not fully able to answer. The array of experiments we studied included virtually no instances of systematic and organised end-user involvement. There may be several factors at play here, including:

1. The relatively small size of our sample of cases, and the preponderance of relatively small-scale and short-term cases.
2. The existence, as we noted in Chapter 5, of various methods of representing the user other than direct user involvement (e.g. market research, panels of potential users; the I-methodology based on imaginative constructions by engineers etc.); the cost and difficulty of users involvement and the difficulties in integrating actual/potential final-users into digital experiments may make working with other knowledge sources, such as complementary suppliers and intermediate users a more feasible and cost-effective option.
3. The difficulties that may surround end-user involvement at the earliest stages of innovation. Experiences with user involvement in relation to industrial ICT applications show that such naïve users may lack the technical skills and confidence to articulate their requirements effectively (Ehn 1988). Non-specialist users will often see technological determination where it does not exist (for example, seeing the particular features of designed artefacts as necessary and immutable) and tend to defer to technical specialists. They are not accustomed to 'technological fantasy' – developing visions of how technologies might be deployed in their activities.
4. Some kinds of user information may be collected by the private sector players involved, but are likely to be kept commercially confidential, and are unlikely to be shared with other collaborators in the kind of non-commercial collaborative experiments that predominated in our sample of cases. Though ICT suppliers have bee criticised for their reliance on rather rudimentary methods for representing users in the past, this is changing (Cawson, Haddon and Miles 1995, Nicoll 2000). Commercial suppliers, of course, are increasingly opting to conduct market research and acceptability testing on their products as they come near to market. However, many of the methods for tapping user experience are costly.

They are only likely to be feasible where large-scale commercial investments are being made. These methods in turn yield market related information which is potentially enormously valuable for the commercial players involved. However, the firms are likely to treat their own market research as proprietary – commercially in confidence – and are unlikely to share this information. Sharing access to the user has been experienced as a particular problem (see, for example, Nicoll 2000), as firms are likely to keep secret their proprietary knowledges of and linkages with their own customer base. Equally, firms are unlikely to place this valuable knowledge in the semi-public domain of a digital experiment.

In many of the digital experiments we studied the involvement of **proxies for the final-user** remained the dominant approach – reflecting the costs and difficulties of directly involving users. This was particularly the case where there is a significant technical development effort (i.e. where the 'mode of control' prevailed over the 'mode of experimentation').

We note further that the various methods of acquiring knowledge of the user and/or involving the user provide greater or lesser opportunities for the user to become involved as an actor. In some cases the user is largely passive and primarily an object of study (Nicoll 2000). However, effective user participation may depend upon the users acquiring the skills and confidence needed to become actors – systems builders/intermediaries – in their own right (Jaeger and Qvortrup 1991, Jaeger, Slack and Williams, 2000).

Another aspect of this issue concerns the asymmetry between final- (or end-) users and other players more closely involved in technology supply, which bears upon their motivation and willingness to be involved in a development project. The core purposes of final-users (for example organisations using ICT applications) may not be strongly linked to technology per se. Instead their interest revolves around the extent to which they may be able to appropriate ICT applications for their particular purposes. In other words, their interest is likely to be more contingent, and limited to a particular project. Their position stands in contrast to players more closely tied to technology supply, who are more likely to find it in either their individual career interests or their organisation's interest to participate in various projects over time (see also our earlier discussion of the role of intermediaries).

One development that is particularly interesting here is the success of ICT suppliers in enlisting the help of users in identifying problems in the technical performance and ease of use of their offerings at an early stage by setting up alpha and beta testing of their offerings. Many of those involved in such testing will, of course, be intermediate users of generic technologies who may anticipate benefits from having early access to new products (coupled perhaps with the expected status and credibility benefits of being a member of this 'club'). However, alpha and beta testing extends remarkably far into the community of individual users of technology (including technology firms and professionals and enthusiastic amateurs) who seem willing to give rather freely of their efforts in exchange for very modest material incentives (there are of course important knowledge benefits including early access to emerging technologies). The phenomenon of 'early adopters' of technology is well known. We should, however, bear in mind that such a technology enthusiast culture may represent a poor basis for reliable extrapolation about the potential wider market future users. ICTs could provide an important medium for feedback from the widely dispersed and anonymous customers of mass-market ICT products. However, outside 'technology enthusiast' cultures, such avenues are rudimentary, though they can be expected to increase. Perhaps the best example from our research was Girls ROM, where the designer of this CD-ROM product set up a comment facility on its associated web pages to receive input on the design of next version. Sørensen and Spilker (1999) describe this as a 'weak' form of the learning economy because, although there is interaction between design and use, this is 'electronically mediated, accidental in nature and separated in time and space'.[1] (Sørensen and Spilker 1999: 71)

It seems that these forms of distributed social learning are prevailing over the direct representation of end-users. This confirms our earlier criticisms of the design fallacy underpinning many concepts of prior user involvement, and points instead to the effectiveness of incremental learning across multiple overlapping cycles of product development and appropriation.

Modes of design and user representation

As well as finding differences in the extent and manner of user involvement, our empirical work identified rather differing conceptions of the role of the designer and the way in which they represented the user in design. This observation emerged when we began to make comparisons between the different integrated study areas, as well as examining differences between cases within domains. In particular we found important differences (alongside a number of similarities) between the cultural content products and our other integrated study areas (education and digital cities) in terms of the manner in which design (and in particular user representation) was approached. To capture these we found it useful to draw a distinction between two possible modes of design focusing respectively on accountability and creativity:

1. *Accountable design*: this mode is rooted in computer system design traditions and stresses the *accountability* of the designer to the user. It is associated with an inclusive approach to design, targeted towards the full range of different user constituencies. In this model, good design is design which meets user specifications – a model which perhaps reaches its zenith in Structured Systems Design Methodologies which formalise the process of drawing up the requirements specification and their signing off by the user/client. The ultimate objective of such a model is a system that fully meets the established needs of a set of identified classes of users. In the rhetorics of accountable design, the technical expertise is subordinated to the user's requirements. This Accountable Design mode has, arguably, constituted the dominant discourse around computer-systems design.
2. *Creative design:* this approach is underpinned by rather different models of the process of design and the role of the design specialist draws upon the rather different traditions associated especially with various kinds of artistic activity – particularly fine art and literature, but also the mass media, and, in a different domain, architecture – and sees design in terms of *creativity,* and valorises *authorship*. In this mode the designer is given leave, based on internalised representations of users, to construct new concepts of use – to reconstruct the user – within the bounds of what particular users can be convinced is acceptable/attractive. Indeed, creativity may be seen in terms of 'moving the user on', and transforming existing genres. The ideal-type is a system that attracts, engages and enchants the user.

The creative design mode would seem to be more typical of many kinds of cultural products – e.g. the Nerve Centre or games design. Here the crucial skill-requirements were perhaps less to do with creating a functioning ICT system, than in creating novel and attractive ways of presenting and organising information. Journalists, graphic artists and musicians seemed to play a key role, as do technical specialists, and bring rather different concepts of the design process. This approach also seems to apply, for example, to DDS, where a group of technologists and other enthusiasts (e.g. local media) shared a particular agenda and vision of how ICT might be used to underpin community information exchange. This creative attempt to enchant and attract users by artistic and technology enthusiast sub-cultures may present a more effective way of proactively eliciting user desires and interest, of anticipating and shaping emerging uses and user needs than the reactive feedback from the user implied by the conception of system design as *accountability*. In contrast the accountability model is more representative of the mainstream of systems engineering, explicitly in the area of participatory system design, but also implicitly in contemporary rhetorics surrounding the business application of ICT 'fitting technology to business and organisational needs'.[2]

The dichotomy between creativity and accountability may be associated with rather different temporal and spatial conceptions of how the user is to be represented – at least at the starting point of design. In creativity the user is primarily represented by an internalised model within the designer's imagination and drawing on the design sub-culture. The ideal-type of accountable design might start, in contrast, with a formal statement of user requirements, derived from the user.[3] There have recently been attempts to make the education of software designers more like, for example, in architectural design as a way of encouraging technical specialists to take creative leaps and then articulate how the designed functions may meet user needs (Kuhn 1996).

MANAGING THE RELATIONSHIP BETWEEN THE PROJECT AND ITS CONTEXT

Learning by regulating: configuring the translation terrain

In a digital experiment, not only must the internal context be managed, but also the relationship between the project and its context. Our empirical work addressed not just the 'internal' issues arising and lessons learnt regarding the conduct of digital experiments, but also the broader processes of learning by regulating, whereby local actors sought to influence other players around them, shaping policy and the orientations of commercial players, to create a context that would be more favourable to their particular projects and commitments. In this section we are, inevitably, also discussing 'what intermediaries do' in promoting and establishing digital experiments. This kind of 'learning-by-interacting' involves two interpenetrating aspects:

- The move from the local to the global – the dissemination, of particular local experiences, perhaps as a currency of success stories or best practice models.
- The move from the global to the local as general resources (best practice models etc.) are selected, adapted and fitted to the particularities of the local context.

Configuring the global; configuring the local

We saw these kinds of process at play in our earlier discussion (see Chapter 5) of the resort to best practice models. The clearest examples arose in relation to digital cities, where there was a relatively organised process of promulgation of particular projects as exemplars, and frequent resort to such exemplars by proponents of a particular local initiative to legitimate the project and particular design choices. The availability of such an exemplar could be an important resource for proponents of such a project – a resource which, importantly, may have a currency across the boundaries within and between organisations. Exemplars can be taken as a demonstration of the potential for such a project and some kind of assurance about its performativity. In a context of uncertainty best practice models provide some kind of evidence of likely success. Mimicry is thus a strategy for wider uptake. However, this kind of learning by interacting does not necessarily impose uniformity in terms of which exemplars are drawn upon or how they are implemented. It is important also to show that the selected exemplar is an appropriate model in a particular, and perhaps rather different, context. This can be a contentious matter, as we saw in the case of Frihuus, where questions were retrospectively raised about the selection of the Salford GEMSIS 2000 project as exemplar for a community information system in Fredrikstad (Brosveet 1998).

Although many of the technical and intellectual resources available for building ICT applications are, at least in some respects, globally available, the projects that result are to a significant extent unique, configured around a particular array of local conditions. Digital experiments take place in diverse settings. Moreover, the local inception and implementation of a project involves a complex set of interactions amongst an array of actors and factors. And the outcomes of this project configuration process are consequently diverse and difficult to predict. Many projects involved collaborations between diverse players with different commitments and expertise (as already noted in our discussion of intermediation as often a multi-level game). We analysed these settings as constituting particular 'translation terrains' shaping development. Our case studies highlighted the contingent nature of many of the problems addressed and the creative way these were tackled by intermediaries.

We saw the radical translation between the original conception and ultimate outcome in the case of the Craigmillar Community Information Service. This was initiated by a desire amongst civil servants and Edinburgh Council officials that Scotland should launch a project to emulate the Manchester Host. Craigmillar emerged from the search for a suitable candidate community where the central idea of using ICT to integrate communities could be applied, as an economically deprived area with numerous welfare groups. However, the process necessitated a reconfiguration and translation of this vision. In particular a much lower funding base first suggested a local rather than a city-wide initiative and then underpinned the idea of developing a system that linked the existing (and already purchased) computer systems of the various welfare groups within the community.

A number of contingent decisions had to be made in the development of the Language Course project, which was motivated by the desire of an equipment supplier to demonstrate its capability (and the subsidiary element – teleteaching Dutch language matched perceived problems in the provision of minority language teaching), and other elements were configured around this constraint (above all the need for rapid development to meet a forthcoming industry exhibition) (Mourik 2001).

Case Summary 7.2
Language Course: technology and education as marketing strategy
This case describes the attempts of a commercially operated big telecom player to enter and explore new market segments. The company's desire to flag its capabilities in new telecommunications markets combined with its Marketing Communications Division's (MCDds) need to demonstrate to headquarters that they could do more than just traditional telephony. MCD sought some stunts to demonstrate the promise of technology. They decided *i.a.* that a distance education with children would have the best effects in terms of public exposure (a European-wide language course was seen as a suitable candidate). So, at short notice, a project was scheduled that would show the company to be a player in the relatively new field of educational technology. An experiment was designed that incorporated the technological promise of using ISDN-facilities in an experimental set-up that connected ten children to one single control unit. The video-conferencing facility encountered technical problems (e.g. communications failures) and a lack of understanding of the user setting – including a failure to anticipate opportunities for playful use of the system by the children (e.g. voice switching of the camera had to be stopped to prevent accidental or mischievous triggering).

In this case, educational objectives were sacrificed for the sake of meeting the deadline of a technology fair, in which the result of the project would be officially presented. Though the project might be termed a failure from many perspectives (be it technological or educational) MCD and the company were very satisfied with the public image results of the project, attesting to the negotiability of attributions of success (Mourik 2001).

Some similar elements came up in Cable School, which was promoted by Cable Co. as a way of demonstrating its capability to the local authority in the run-up to anticipated competitive tendering for provision of a new network. Figure 7.4 draws attention to the lack of fit between project development constituency and social constitution of the schools: between vertically structured project management structures (the networks or players involved in planning, managing and carrying forward the Cable School project) and horizontal communities of teachers and students in the school, who remained largely excluded from project planning and management.

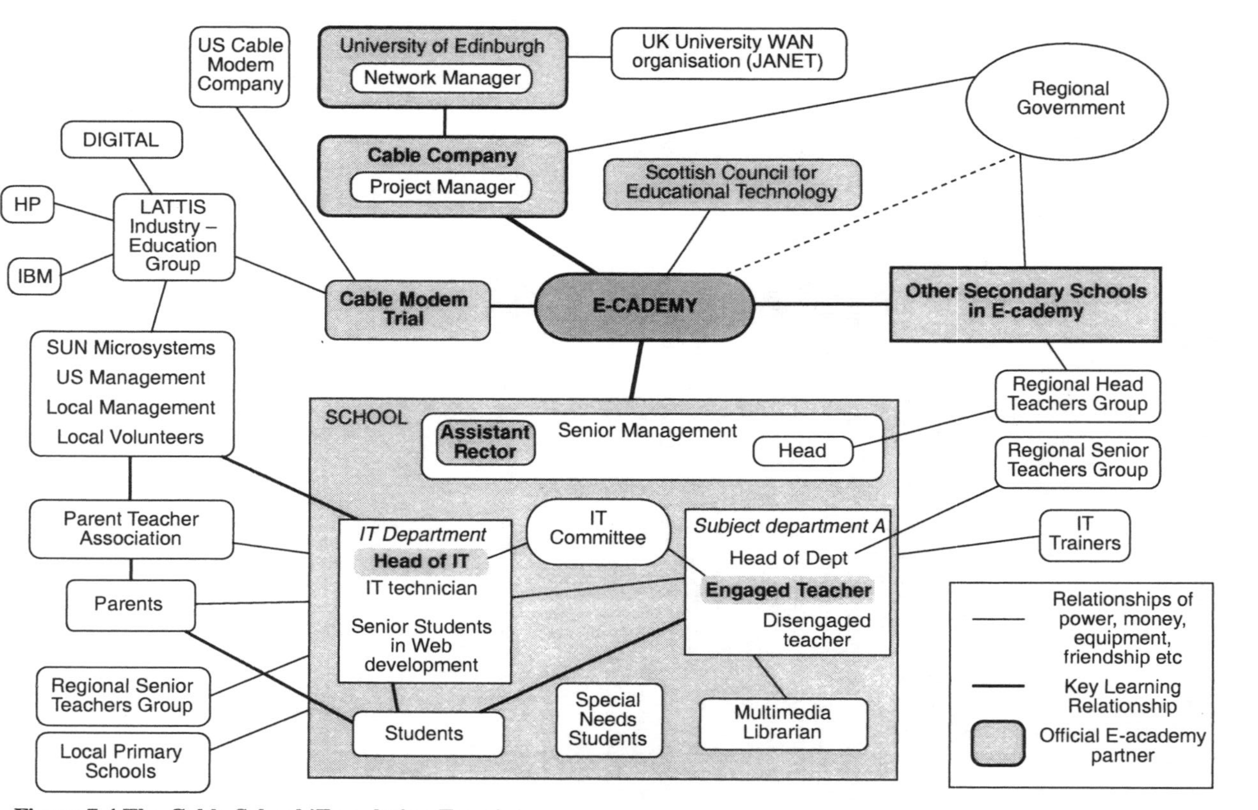

Figure 7.4 The Cable School 'Translation Terrain'

NOTES

1. Thus Sørensen and Spilker note (Sørensen and Spilker 1999: section 2.3): 'the home seems to be weakly linked to the major learning economies related to the ICT industry. Even if the Internet allows new opportunities for users and producers to communicate in an informal way through web pages and email, it is unclear whether these opportunities are made use of. In fact, one may suspect that most companies consider their home pages to be sites of advertising, information and downloading of updates, rather than as a real medium of communication with their customers/users. Social learning of ICT in households may be embedded in small user-to-user learning economies, but these economies do not appear to be linked to user-producer networks in any substantial manner. This means that producers miss out on opportunities to 'tap' core user groups for ideas, information and experience, while users do not get the information and advice from designers that they could have given them substantial benefits.'
2. Paradoxically, design as creativity, despite its emphasis on openness of outcomes, resembles the Woolgarian model of design 'configuring the user'.
3. Making such a dichotomy between creativity and accountability of course immediately draws our attention to similarities between cultural and other products. For example, Hollywood films and computer games are now put through increasingly formalised market testing before release.

PART B

Rethinking Innovation Models and Technology Policy Perspectives

8. Policy Contexts, and Debates: National Settings for ICT Adoption

In Part B of this book we step back from empirically-driven analysis of the processes on innovation and social learning around the development and use of technologies to explore the broader implications of these findings for public policymakers, corporate decision-makers and others seeking to influence the process of technological innovation in ICT. However, to convey this policy contribution effectively it is first necessary to explain something of the policy contexts of the study.

Chapter 8 seeks to capture the historical policy context and the concerns and debates that have informed this enquiry. We start this chapter by reviewing, and analysing critically, some of the prevailing presumptions about the relationship between technical change and social change with ICTs. We focus in particular on the widespread 'received view' underpinning ideas of digital revolution and the transition to an information society – ideas which are strongly influenced by technological-determinism, and the rhetorics of technology supply. We need to convey this simplistic and flawed vision because it constituted a point of departure: in so far as the social learning perspective emerged in part through an attempt to combat these ideas.

The rest of Chapter 8 reports on a series of studies we undertook at the very start of this investigation across nine European countries of the national context – industrial, societal and policy – for ICT development and appropriation.

Our main findings are analysed in Chapter 9, which, by way of a conclusion to the whole book, considers some of the policy implications that have emerged from our theoretical explication of social learning and from the substantial body of empirical research and analysis reported in Part A.

TACKLING 'RECEIVED VIEWS' OF THE DIGITAL REVOLUTION AND INFORMATION SOCIETY

It has been widely presumed that 'multimedia' digital data-processing systems that could handle text, graphics, sound and video signals would be easier to use and more engaging for consumers. The emerging 'information superhighways' would bring these ICT products and services into every home and workplace. By the mid-1990s, public policy and debate focused upon the implications of these changes – with new policy frameworks emerging from the USA, Japan, Europe and its member states (Kubicek, Dutton and Williams 1997). The influential Bangemann Report (High-Level Group on the information society 1994) argued that Europe was becoming an Information society – already in the throes of a new industrial revolution 'as significant and far-reaching as those of the past'. The Bangemann Group saw ICTs as leading to new jobs and wealth creation, coupled with 'a significantly higher quality of life and a wider choice of services and entertainment'. (idem: 10)

The Bangemann Report was extremely influential, especially within Europe. It forms part of a broader current of discourse about the imputed *digital revolution* and the shape of the emerging Information Society. Though there are important differences between particular accounts, many of these discourses share some broadly common elements, in particular in relation to their view of technology as a driver of social transformation. We will briefly review these suppositions, taking Bangemann as an exemplar, as they provide something of the intellectual context for this study, constituting a set of counter-theses with which our project sought to engage critically.

The Bangemann Report argued that ICTs posed a 'revolutionary challenge to decision-makers': a revolution that would generate 'uncertainty, discontinuity and opportunity' (High-Level Group on the Information society 1994: 10). The key technical change in question was the imputed technologically driven convergence: the coming together of core data transmission and processing technologies into an integrated digitised technology infrastructure able to handle different kinds of signal (data, voice, graphics, video); the resulting convergence of computing, communications and broadcasting; and, finally, the integration of different industries, services and realms of activity through these digitised media (Forester 1985). This kind of presumption of a shift in 'techno-economic paradigm' (Freeman, Clarke and Soete 1982, Perez 1983) tends to be conceived as a change whose broad directions are in some ways already mapped out. They take as their starting point the perspectives and claims of technology supply (and as a result we have referred to them by the shorthand 'the rhetorics of technology supply'). From such a view, the main risk may be of 'missing the wave' of change or coming late to the revolution; as a result the key policy questions concern, on the one hand, how to increase the pace of technological change and, on the other, how quickly the socio-economic system can adapt to the new exigencies of technological advance. These kinds of view have been criticised for their 'technological determinism' – for conceiving social change as being driven by changes in core technologies. This kind of approach is readily exemplified in Negroponte's *Being Digital*, which argues that innovation could be explained 'through purely technological imperatives' (1995: 81). Subsequent national and European policies have moved away from this narrow deterministic focus. However, this kind of deterministic discourse has been remarkably resilient, particularly within policy circles, and it is therefore important to explore these views and subject them to critical analysis. We will first examine some of their common presumptions.

These deterministic discourses, of revolutionary social and institutional change, driven by technological advance, tend to embody a rather monolithic concept of technological development. By implication, the specific *technological paths and choices are not seen as problematic*, as if the *trajectories of technology* were well established. Finally, ICT is treated as if it already provided ready *solutions to social need.* A further consequence is that they tend to emphasise the *global* character of these changes. We shall examine each of these presumptions in turn.

ICT development paths and trajectories are unproblematic

In this kind of global and monolithic account, the particular development paths and choices for the application of ICT are not seen as problematic. In some accounts this reflects a presumption that there is a well-established development trajectory – typically revolving around imputed trends in core technologies (for example, ideas about the convergence of currently separate media and services [broadcasting, telephony, computing] into a single digitised medium). Other accounts see 'the market' as the driving force. The Bangemann Report falls in to the latter category. Its emphatically *laissez-faire* view insists that the market, rather than the state, will 'decide winners and losers'. The 'prime task of government is to safeguard competitive forces and ensure a strong and lasting political welcome for the information society, so that demand-pull can finance growth' (High-Level Group on the Information Society 1994: 13).

However, at the time that this study was being conceived, the long-awaited 'killer-application' that was expected to bring every home on-line had still not emerged (Dutton 1995, 1996). Some features became established in the course of the research: for example, the exponential growth in the Internet and World Wide Web have consolidated their position as probably the dominant framework for on-line services. Digital TV and other platforms (notably mobile telephony) were and still are 'waiting in the wings' however. A variety of technical configurations are possible. We will suggest, below, that the outcome is likely to be a web of overlapping, rather technologically heterogeneous, webs rather than the kind of orderly and seamless web suggested by 'information superhighway' concepts). There is, moreover, little certainty over what kinds of service model will prevail.

ICT offers already available solutions to social need – that merely need to be diffused into society

The presumption underlying Bangemann's laissez-faire view is that ICT is already available, more or less off the shelf, and that it 'offers attractive and cost-efficient solutions to meet consumer needs' (European Commission COM 1996a: 3). In this case, the key remaining task is to ensure the wider diffusion of these technologies and the growth of markets for ICT products and services.

Somewhat paradoxically, in this connection, one of the most important features of the Bangemann Report is its proposal to launch *experimental applications* of ICT, motivated by recognition of the possibility of market failure where a technology cannot establish the critical mass of users to become financially viable. While emphasising the importance of market forces and competition, Bangemann notes that this by itself will fail to 'produce – or produce too slowly – the critical mass which has the power to drive investment in new networks and services'. In order to 'create a virtuous circle of supply and demand' it calls for a significant number of market testing applications to be launched across Europe to create critical mass.

> Initiatives taking the form of experimental applications are the most effective means of addressing the slow take-off of demand and supply. They have a demonstration function which would help to promote their wider use; they provide an early test bed for suppliers to fine-tune applications to customer requirements, and they can stimulate advanced users, still relatively few in number in Europe as compared to the US. (High-Level Group on the Information Society 1994: 27).

In making these recommendations, Bangemann insisted that these initiatives 'are not pilot projects in the traditional sense' and but are demonstrators: since their first objective is for market testing 'to test the value to the user, and the economic feasibility of the information systems'. (idem: 28) They therefore need to be launched in real commercial environments, preferably on a large scale to be truly effective. Despite this insistence on the limited role for public initiatives, public support given to public and private experiments has proved to be one of the most important and effective outcomes to emerge following the Bangemann report.[1]

The value of these initiatives has not been as narrow demonstrators of a finished technology, but as a space for experimenting and learning, bringing together technology providers with other players (e.g. service providers and some kinds of users) in addressing not only technical and operational aspects but also questions about usage and user responses. Initiatives designed as demonstrators provide important opportunities for broader and more open-ended learning.

Our research project came to the view that we could take this observation further to suggest that all ICT projects are in some ways experimental – not just those explicitly established as feasibility studies or social experiments. However, one of the main conclusions to emerge from this study is that the failure to recognise the experimental character of ICT initiatives means that the lessons from these projects are often not sought or properly utilised. ICT offerings are far from being 'finished products' in a number of ways:

1. First the technologies are still unstable. They are still evolving rapidly and often prove rather difficult to deploy.
2. More profoundly their applications and uses are, with only a very few exceptions, still a long way away from becoming established and widely accepted. There is still a need for inventive activity – by suppliers, by producers of content and by users – to work out ways in which these technologies may be made relevant and useful to potential consumers.
3. This lack of stabilisation and 'closure' of technology is compounded in the case of ICTs by the fact that these are designed to be flexible technologies – able to be reconfigured and applied in ways not anticipated by the designers. They increasingly take the form of generic platforms and media that can be used with a range of applications, services and types of information and cultural content. In this sense ICT applications may never become 'finished products', demonstrating product stabilisation and maturation of markets in the way that has been seen with other consumer goods.

These observations lead us to move away from conventional technology policy approaches which conceive innovation as limited to initial research and development and followed by a separate stage of technology diffusion (for example through demonstrator projects and awareness campaigns). Instead, a rather different kind of activity may be needed that recognises the important innovative effort that follows initial product design and development and focuses upon the local *appropriation* of generic ICT capabilities. We return to this in the next section.

Technology as a global development

These ICT innovations are portrayed as part of a set of global developments. Technological capabilities are seen to have global applicability, and a global market is presumed for new technology products and services. One key question that derives from this view concerns what share of this global trade

Europe shall obtain. Under such a view, the main question about our future concerns which countries will 'get there first' and thereby benefit from new industries that will emerge. Thus, Bangemann insisted if 'Europe arrives late our suppliers of technologies and services will lack the commercial muscle to win a share of the enormous global opportunities which lie ahead' (High-Level Group on the Information Society 1994: 12). This conception of the global ICT market – with the major economic rewards going overwhelmingly to the successful globally dominant players – sets up an extremely pessimistic view of Europe's prospects given the current strength of the USA in ICT supply and in the Internet. Such a view could be damaging to the extent that it understates the potential opportunities for European firms by concentrating, as it does, on the most globalised markets (e.g. the core technologies underlying computer platforms and delivery systems) and downplaying the much larger market for products and services (e.g. business and information services and cultural content) which have a much stronger local dimension.

This globalised and monolithic concept of ICTs is often seen to carry a further implication that the global availability of technological capabilities, and the increasing role of global players in technology and service supply will result in the progressive homogenisation of technical infrastructures, applications and service concepts – and, through this, a homogenisation of societies as a whole.

The question of whether ICT is indeed an international medium that cuts across and erodes national differences or whether it is appropriated differently within particular local settings is perhaps the central issue addressed in our survey (reported below) of national patterns of ICT uptake. One of the main findings concerns the enormous diversity of models of ICT application between countries (and, for that matter, within countries), given their different histories, their differing industrial, geographic, technical and cultural contexts. Generic technological capabilities can be taken up differently in different settings. The local dimension proves even more significant in relation to ICT applications and the services, information and cultural content that run upon them.

This finding has important socio-economic implications. Though some elements of the Information society may be globally available, others will remain localised. Local initiatives will continue to be key. Technology policy needs to encompass this – and some of the main recommendations emerging from this study concern the importance of local appropriation efforts for ICT services and content. This finding brings similar economic implications that some of the key market opportunities may be for local applications. We will suggest later that the scope for local markets is particularly marked in relation to cultural products with local symbolism – for example those concerned with local identities.

GOALS OF THE STUDY OF NATIONAL SETTINGS FOR ICT APPROPRIATION

Cross-national comparison and international research offers an important contribution to understanding the social shaping of ICT and the scope for social learning. Specifically it opens up opportunities to examine diversity in the way ICT has been conceived; the influence of particular national and regional socio-economic settings in shaping these concepts; and the interplay between global and local factors in this process.

One of the key questions relating to the societal appropriation of ICT concerns whether ICT is an *international medium* that cuts across and erodes national differences or whether it is *appropriated differently* within particular national techno-economic and cultural contexts? Modern technologies are often conceived as providing universally applicable techniques and solutions that can be applied in a wide range of social settings; the use of technologies is seen as a simple extension of their powerful technical functionality (often taken as self-evident). Such technological determinist discourses have tended to portray ICT as a vehicle for global trends and indeed have proposed some kind of societal convergence. On the other hand, more sociologically informed accounts have tended to emphasise the contingencies and specificities that surround the application and consumption of technology – and in particular the enduring importance of local cultures and actors. Empirical research (and common sense) would suggest that global and local are both important, and that what is at issue is the interplay and balance between the global and the local in ICT development and use (Castells, 1996).

Here the social learning perspective suggests that the uses and outcomes of technologies are not simply inscribed in the design of artefacts, but depend equally upon the way these are selectively taken up, adapted and incorporated within the practices of individuals and groups. And it is in these local consumption processes that the meanings and significance of technologies are ultimately expressed and achieved. As a result, intermediate and final consumers are part of the equation as well as technology promoters and suppliers. This is particularly marked in relation to technologies such as ICT, which are not restricted to particular industrial or occupational settings, but which are seen as becoming widely adopted across all areas of work and everyday life.

Some of the elements of ICT are indeed designed and produced at a global level. Certain players have achieved global reach – for example microprocessor manufacturers, who must pursue enormous economies of scale from global production to cover the escalating costs of developing new products. However, the development and implementation of ICT delivery systems and applications must inevitably include a local element. The importance of the local may vary, for example, between simply adopting and configuring standard commodified products to the wholly customised production of products and services for particular national or regional users or niche markets. There may be significant differences between different parts of the ICT field. The core hardware and software technologies at the heart of the information infrastructure may, of course, largely conform to global technical standards; their selection may be strongly shaped by widely established 'technical' criteria of price and performance. On the other hand, the applications and content that run upon this infrastructure are more liable to be specific to particular local cultures, contexts and requirements. Important differences can be noted in this respect between sectors and application domains. For example, one may expect to find more international standardisation in relation to on-line banking or credit-card services than for cultural products; the former two are perhaps more subject to common functional requirements and there are strong pressures and templates for international harmonisation.

The balance between local and global in the societal appropriation of ICT is thus likely to vary. This has important economic, social and policy implications. For example, the respective contribution of local and global players in ICT provision bears critically upon the opportunities for national and regional economies to become involved. This balance of contributions is neither predetermined nor uniform between countries and sectors, but will depend upon which players are active, their resources, the relationships between them (and their relationship with global players) and the particular context (especially the market size, but also government policies etc.).

There is an international community of ICT suppliers and application developers, interacting tacitly and sometimes explicitly (e.g. through various international fora) to create common visions of the utility and benefits of new products and services. However, even these global visions – for example of e-commerce, of ICT in health, education and public administration – must be implemented in specific national, regional and organisational contexts. National or regional settings present more or less distinctive terrains for the appropriation of ICT, shaping the strategies by which local players (and global players acting locally) may seek to develop and implement ICT applications, This raises a key research question for the SLIM study concerning how national and regional settings shape the ways in which opportunities for ICT application are conceived and pursued. This is the topic which lies at the heart of this part of the SLIM research project.

To explore these questions, the SLIM research network embarked upon a cross-national investigation through a series of national studies geared towards comparative insight. These sought to understand the national setting for the appropriation of ICT application, how this might be patterned by structural (e.g. demographic) and institutional factors and how this, in turn, might influence the strategies and visions of the main players involved. These national studies also included a review of social experiments and trials in ICT, conducted as a prelude to a series of detailed case studies of the development and use of ICT applications that formed the main empirical focus of the SLIM project.

METHODOLOGY AND CONDUCT OF NATIONAL STUDIES

Methodology for national studies

The national studies sought to provide a broad coverage of a set of extremely complex and rapidly changing developments – and moreover undertake this in a short period and with a very limited level of resources for each country. We were not therefore able to undertake a large-scale survey, *ab initio*, that could be based upon a common methodology. Instead it was necessary to rely, where these were available, upon existing national statistics, published material and the wealth of 'grey literature' in this field for each country, supplemented by selected interviews with key players. This inevitably entailed a number of problems with the interpretation of material – and necessitates particular care in relation to making detailed comparison.[2] In particular, reliance on diverse (and extremely variable) secondary sources meant that it was not possible to adopt a wholly standardised methodology for the national studies.[3]

To aid in the conduct of the studies and presentation of their findings guidelines were agreed amongst the SLIM research teams regarding the methodology, content and format of the national studies. The national studies covered Belgium, Denmark, France, Germany, Ireland, Japan, the Netherlands, Norway, Switzerland and the UK.[4] They have been published separately by Williams and Slack (1999).

Review of digital experiments

Each national study included a review of social experiments and trials. For this part of the study, a more rigorous methodological foundation was possible, supplementing published sources such as official documents with interviews with key players. These reviews were not only useful in their own right, but also provided the initial sample from which to draw the set of more detailed 'Integrated Studies' of the development and uptake of ICT products and services in Education, Public Administration and Cultural Products.

International developments

Though not able to investigate global ICT developments per se the SLIM programme also explored how such developments, such as international standard setting and the activities of international suppliers, impinged upon ICT development/appropriation strategies within Europe.[5]

Analysis of the national studies

The national studies sought to capture how the national/regional setting and the array of major actors patterned the dominant conceptions of ICT and the perceived 'rules of the game' for ICT innovation. The descriptive studies and their analysis were 'brought to life' by interrogating the available sources regarding a variety of research hypotheses about how the national/regional setting might shape the appropriation of ICT. These hypotheses served as a sensitisation tool for the researchers in conducting their national study, and guided the comparative analysis of findings. These hypotheses can be grouped around three areas:

1. the influence of the character and orientation of *the main ICT actors and 'constituencies'* in the nation.
2. the influence of *the demographic, geographical and cultural context.*
3. the influence of *the policy context.*

We shall examine each of these in turn in the three sub-sections that follow. They are however closely related. Thus, as we see, public policy and regulation influence the salience of national players – notably the number and role of telecommunications providers. Our subsequent analysis conversely highlights the influence of national industrial players on public policy.

MAIN ICT ACTORS AND CONSTITUENCIES

Each SLIM research centre undertook a study of how national/regional settings shaped the prevailing ways in which ICT was conceived, and the strategies for the appropriation of ICT of the main actors involved.

An important empirical focus of the study was thus upon the structure and operation of the **ICT constituencies** in their national/regional settings. This addressed the identity and strategies of the key players involved in ICT provision, the relationships that existed between them and with external players. It sought to explore how the structure and orientation of these national ICT constituencies shaped the ways in which opportunities for ICT are conceived and the overall pattern of ICT developments. We were interested here in assessing how the strategies of these various players and the interactions between them set the 'rules of the game' as well as the attendant models and visions of ICT and its uses – in this way perhaps constituting a particular 'technology regime' (Rip 1995).

The starting point was an exercise to map the ICT constituency, by identifying the main active players – who constituted the 'key drivers' within the nation/region. These players potentially included research organisations and relevant public policy bodies, telecommunication service providers, and the range of firms supplying ICT and complementary products and services – 'intermediate users' such as media and business information companies.[6] The study sought to address the relationships between the main players (e.g. ownership, collaboration) and their relationship with global players. It examined the economic strength and technological competencies of these players, and other factors bearing upon their influence (an important question here concerned telecommunication service providers; whether they were in public or private ownership, and the level of competition in telecommunications provision – see below). Research sought to establish which concepts of ICT these key players promote, and how this was shaping overall conceptions of ICT. We were particularly interested to examine the concepts they had adopted of the market for ICT products and services – particularly regarding who they see as their user and what uses they envisage.

An important variable is the extent to which there are 'national champions' in ICT provision in particular nations – and in this case, whether and how that player influenced the conceptions of ICT that prevailed? Such a 'champion' might be a major ICT company or the national telecommunications provider. Only a few countries have major ICT suppliers (Siemens in Germany; Philips in the Netherlands, Thomson in France); all countries have powerful established telecommunications companies – Telcos – though their relative strength and roles may vary. Other players, in particular from existing media industries, could also be significant (notably in Germany and the UK). The national, and indeed global, strategies of these players could be a significant influence over ICT initiatives, public policies and wider perceptions of technology within a nation – as already noted in relation to the salience given to digital TV in Germany.

Our concern to address how the industrial base and expertise within a nation (including the presence or absence of a national champion) shape the way ICT is pursued required some assessment of the respective strengths of the main players or groups of players – such as telecoms network providers, the TV/video producers, cable companies, the press etc. – as well as the relationships that existed between them.

Finally, the studies sought to address how the ICT constituency is changing – with the emergence of new actors and the evolution of existing actors, together with shifts in their respective roles and power.

Regulation and competition in telecommunications

Established national telecommunications providers are some of the largest technology players. One crucial feature in relation to the operation of national ICT constituencies and policy formation (see below) appeared to be the level of competition in telecommunications (and Value Added Network Service [VANS]) provision. There is a global trend, underpinned by the policies of the World Trade Organisation and the European Union towards 'liberalisation' – and in particular the privatisation of Telcos and the introduction of competition in telecommunications provision. However, there were marked differences across the European countries studied by SLIM. On the one hand, were countries like the UK and Norway, which had made early and rapid moves towards liberalisation, and at the other extreme France, Germany and Ireland in which the process is still underway.

The extent of liberalisation in telecommunications services – the level of competition, i.e. whether there was a single dominant player or multiple telecoms providers, together with the regulatory restrictions that might apply to the activities of players – could have important consequences for the development of networked ICT trials and services. This was particularly an issue in the aftermath of liberalisation in which the incumbent national postal, telegraph and telephone (PTT) would typically start from a monopoly position. In this context, public policy and regulation might be driven by the need to offset tendencies towards monopoly and the exclusion of potential entrants by placing restrictions on dominant players. One example can be provided by the UK, in which industrial competition policy constrained the pricing policies of BT, the former public monopoly provider, and protected new entrants to the telephony market, while also preventing it from providing local video-services to give time for the fledgling cable franchises to become established (Stewart and Slack 1999).[7]

Where there was a single dominant supplier (i.e. a national monopoly or effective monopoly), that player would be virtually a necessary partner in any ICT pilot or initiative. This would also affect the parameters for participation in trials. In a monopoly or near monopoly context a dominant Telco may be motivated to take part simply to help stimulate uptake of telecommunications services overall – thereby building its own market. Such firms were often in a position to make substantial investments (for example Telecom Eirean is a significant player in Ireland [Kerr and Preston 1999] and has recently funded a 'cabled town' initiative). In addition to their large size, their monopoly situation allows them to deploy considerable resources, whilst minimising risk – for example by channelling user responses and shaping market development. In a more competitive context, Telcos may need to give more attention to whether, when and how they will gain a return on investments in such third-party trials, and their attempts to prefigure and stimulate market developments may encounter more immediate challenges. For example, France telecom, which in the past has been a major player in large-scale ICT developments such as Minitel, has become more cautious since telecommunications privatisation and deregulation (Vedel 1996, Lobet-Maris 1999).

Though the extent of progress towards liberalisation of telecommunications was an important shaping factors across a range of countries examined, it would be unhelpful to see this as a uniform process leading to a single set of outcomes. There is certainly a shift (and a deliberate shift) towards plurality in telecommunications provision. However, developments were complex – and particular historical contingencies continued to shape national provision in particular ways. For example, in the Netherlands, for historical reasons, local cable companies have found themselves in a key position in ICT developments (van Lente 1999).

THE DEMOGRAPHIC, GEOGRAPHICAL AND CULTURAL CONTEXT

The research sought to identify how the national appropriation of ICT is shaped by its demographic, geographical and cultural context. These constitute the 'structural' physical and socio-political terrain, providing the setting for other considerations such as public policy and the structure and orientation of ICT constituencies. This part of the study started by examining particular structural features of the national/regional context including, *inter alia*, nation size, geographic and demographic features and linguistic groupings. For example, population size and the size of linguistic groups have a very immediate bearing upon market size – and thus the likely return on investment from developing products for a particular market. We were interested to explore how these factors may have patterned the 'rules of the game' in particular regions.

Let us explore these points in more detail. At the most immediate level, a question arises about how the density and distribution of population affect the use of telematics and other technologies for remote communication. It seems plausible to suggest a relationship between the high level of uptake of cellular telephony in Norway (or for that matter in Finland), the emphasis in that country upon communication and internal linkages in public discourse and policy, and the physical size of the country and physical dispersion of its population. However, the structural context does not per se impose particular views. We can equally find high rates of mobile phone adoption in small, densely populated countries such as Belgium, Ireland, Italy and the UK.

Various factors are clearly at play here, including the commercial and pricing strategies of mobile suppliers (especially regarding the launch of pre-paid phones), and fixed telephone line penetration, and much depends upon how these structural features were construed within social and policy discourses.

The size of linguistic communities provides an interesting illustration of this kind of interplay. One area of interest for the study concerned how countries (or language communities) might react to the dominance of US/English language offerings in the field of ICT. There seemed to be some interesting differences between countries and regions. In some countries there have been strategies to oppose the dominance of US culture and English language products, as part of a broader concern to maintain national cultural integrity.[8] Conversely, there was less evidence of this in certain smaller countries such as Denmark and the Netherlands in which English is already widely spoken. (This may in part be explained by the fact that – for reasons of work, or from the need to follow imported, sub-titled films – many people in these smaller countries have already found it sufficiently attractive/expedient to make the investment needed to learn the English language; in contrast, in larger economies and linguistic groups it may be more cost-effective to support local cultural production or the translation, rather than just sub-titling, of films, and fewer people may be drawn to learn English). The importance of language group and cultural identities is not simply a factor of population size, of course, but relates crucially to matters of history and political and cultural sensibilities. For example, there was more vocal evidence of support for Dutch language programming within the Dutch-speaking communities of Belgium – groups which are concerned to maintain their identity in competition with other linguistic (Francophonic and German) communities – than within the Netherlands itself (van Bastelaer, Lobet-Maris and Pierson 1999). Similarly the most salient initiatives in French language programming were noted in Quebec (van Bastelaer 1999). The importance of cultural and linguistic considerations thus appears to be part of a broader political picture.

THE POLICY CONTEXT

Many countries have developed policies and initiatives about the imputed emergence of national information infrastructures/information super-highways, and the ICT services that would run upon them. These policies are often linked to discourses about an imputed current transition towards an information society. The latest wave of initiatives across Europe and the developed countries seems to have been stimulated by the high-profile launch of the US National Information Infrastructure (NII) initiative in 1993 (Kahin and Wilson 1996), followed in 1994 by the Bangemann Report (the European Union action plan 'Europe's Way to the Information Society') (Schneider 1997). Governments have sought to keep ahead of the field – or at least not be left out of these technological advances, and the social and commercial benefits expected to flow from them through a range of related measures:

1. by creating a supportive policy and regulatory context;
2. by directly or indirectly ensuring the development of the technical infrastructure (e.g. the information superhighways); and,
3. by supporting initiatives and experiments in the application of ICTs.

In their attempts to match or outdo each other, we find some processes of convergence – arising perhaps through mimicry of policies developed in other countries and through alignment of views. Many information society policies thus seem to be united in their resort to arguments about the competitive necessity of being at the forefront of ICT developments, and the dangers of being left behind in the technological competition (van Lente 1999). Another feature has been the articulation of 'technological utopian' views of the presumed societal benefits of these changes (West 1996). On the other hand, differences in emphasis and style remain (e.g. around long-standing differences in national policy styles and contexts [Lobet-Maris 1999]) – for example in the approach to the role of the state and the balance between state and market in overall provision. As we see below, a spectrum of approaches can be found across the European Union with the UK at one end emphasising laissez-faire approaches (Stewart and Slack 1999) and a range of countries attributing a greater role and legitimacy for the state.

To address these, the national studies sought to capture the national policy framework concerning ICT. In addressing this, we were particularly interested to identify whether it was underpinned by a particular conception of ICT, its use and social implications (for example in relation to the dichotomy, discussed below, that was posed between the Internet and digital TV).

In assessing policy frameworks, we sought to identify points of similarity with, and influence by, external reference points and to assess whether governments had adopted these policies wholesale or more selectively. For example, we were interested to assess the influence of the EU Bangemann Report and the activities of the G7 group of countries, which played an active role, for example, in promoting and coordinating digital experiments).

This part of the studies also sought to find some evidence of the consequences of these policies. Did public policy merely shape national rhetorics surrounding ICT, or did it also affect patterns of investment in new ICT infrastructures and service and even the social appropriation of ICT.

A review was undertaken of policy initiatives to promote ICT creation and adoption. This examined the goals of these initiatives, the diverse concerns being pursued around ICT, and the balance between these. This showed that ICT and information society initiatives were geared towards a range of different goals: between economic development goals, for example promoting competitiveness and job creation, and societal goals, for example regulations and policies geared to countering potential negative effects on privacy or social exclusion; fulfilling broader social objectives e.g. for public service and the quality of life or simply the need for democratic control over ICT development. Different parts of the administration – for example, education, trade and industry, local government – tended to have different priorities and concerns. As a result gulfs and contradictions could be noted between the policy perspectives of various public bodies, though the overarching responsibility for policy tended to be located with ministries concerned with trade and industry (Lobet-Maris 1999). We were particularly concerned to examine how this affected the conduct and goals of social experiments and public trials with ICT (see Chapter 4).

The review also sought to characterise the means deployed by the state, which varied across a spectrum, and seemed to be changing. At one extreme was the *dirigiste* model of the state acting as a director of change (selecting and developing particular technologies; providing infrastructures; running projects [and the state is a major potential user of ICT in its many service and administrative activities]). Such a model has deep roots – in interventionist approaches to science and technology, and economic development more generally, and has, for example, been a major feature of the telecommunications technology policy in most European countries until this decade. This tradition has been particularly marked in France, where central-direction over strategic and high-profile technology projects has been given the title 'High Tech Colbertism' – in reference to the successful industrialisation strategy carried out by its eighteenth-century premier (Cohen 1992). However, even here a shift from *dirigisme* to *pragmatism* has been noted, linked to recent failures of large high-technology public policy initiatives, such as the 'cable plan', and the more widespread shift towards liberal approaches (Vedel 1996, Lobet-Maris 1999).[9] In Japan, the state plays a no-less central role, organising choices in innovation and directing the activities of private ICT firms (Collinson 1999). Denmark also stresses the role of the public sector, both as a driver for ICT development and, rather distinctively, as a vehicle for addressing broader social objectives about how ICT will be deployed. Public sector intervention is seen as necessary in Denmark to offset potential negative effects of technology and ensure that it will be geared to particular social objectives which might otherwise be overlooked and also to ensure the involvement of citizens in decision-making over technology (Jaeger and Hansen 1999).

However, a range of less direct models and methods of intervention exist, and extend, at the other end of the spectrum, to a situation where the state merely acts as a coordinator or catalyst of change – for example by facilitating and organising linkages and flows of knowledge between players from industry or technology institutes. In many countries, public intervention around ICT technologies has veered towards the latter form, involving indirect methods, such as training and awareness initiatives, and providing resources (e.g. grants) to help other public or private bodies to carry out projects (Lobet-Maris 1999). Such a model would seem well suited to a setting in which technological decision-making cannot be centralised within the administration. This may arise because choices in technology evolution depend upon the interaction between a range of players – some of which who may operate largely outside the realm of the state's powers (for example global suppliers) or upon the behaviours of widely dispersed consumers which are difficult to predict and organise. As Lobet-Maris (1999) has noted, this is explicitly recognised in the Dutch 'information superhighway' policies are based upon recognition of the limited role a government can play in shaping the Information Society.[10] Similarly, Brosveet and Sørensen (2000) have pointed to the shift in the model of state activity in Norway from planned development and growth of technological capabilities towards a more opportunistic model drawing on global developments, which they characterise as 'Fishing not farming'. Whereas 'farming' describes 1980s strategies of nations to grow technological capacities in core technological fields, 'fishing' implies a selective strategy, drawing on offerings found in that global market. These developments have to be seen in the light of the increasing influence of neo-liberal approaches and the globalisation of technology development.

The language of 'deregulation' should not, however, lead us to confuse this shift with the departure of the state from the field, nor a withering away in the role of the state.[11] Indeed, one of the key findings of this study has been of the central role of public administration (at local, national and international levels) in digital experiments and initiatives (Jaeger, Slack and Williams 2000). Within the European Union, the European Commission has become a major sponsor of trials and projects. Its role is particularly salient in countries such as Ireland which have a modest indigenous development base (Kerr and Preston 1999).

The institutionalisation, structure and traditions of state activities

In carrying out comparisons, an important difference between different European states is in the structure of state activities – particularly regarding the significance of regional and local state structures and in the relationships/allocation of responsibilities between local government, regional structures and national/federal government functions. These differences in governance structures may be significant (and in particular in the extent to which there is a significant regional dimension to state activities), especially since technology development involves both national elements (e.g. the development of information infrastructures) and local elements (e.g. particular applications – such as, for example, digital cities and community information systems). The review therefore sought to capture the involvement of the different levels of government. We were interested to address how the relative salience of different levels of state activity, and the allocation of functions between them might influence the effectiveness of state intervention. For example, might the dilution of responsibilities between different bodies, the lack of funds or power of key bodies; or the strength of laissez-faire ideologies, give rise to weaknesses or gulfs in the state's activity? There is a sense that the federal governance systems in Belgium and Switzerland may have particular problems in developing and implementing technology policies for ICT – for example where responsibility was delegated to regional structures, and overall national coordination or vision were largely lacking (though the lack of a national technology policy in this area is not perceived as problematic within Switzerland) (van Bastelaer, Lobet-Maris and Pierson 1999, Rossel 1999).

Finally, and in part following on from this last observation, we note that the long-standing differences that could be noted between states in their policy concerns and traditions, did not always seem to derive from the specificities of ICT in that country. Not only were existing policy styles reflected within ICT policies, particular national concerns could end up being projected into discussions of ICT policy. For example, concerns about the effectiveness of the Swiss and Belgian federal systems were revisited in their information society debates. Similarly, we note the way in which concerns about inter-communal relations resurfaced in Belgium. In this sense, we could see ICT serving as a kind of 'Rorschach Test' – a recipient of long-standing policy concerns.

The relationship between the state and industry

Any assessment of the significance and impact of state intervention must address the ***reciprocal*** *relationship between the state and industry* (and other technology players such as research and education institutes). On the one hand, the state is an important actor in establishing the terrain and setting the 'rules of the game' for ICT developments (as already illustrated by the important example, already discussed, of the extent of competition in telecommunications). One the other hand, the policies adopted by the state, and the measures available to it for exerting influence, were shaped by the size and structure of the ICT constituencies in these countries. The existence of powerful national champions able to influence the terms of state policy was an important factor. This factor however interacts with the established mode of operation of the state – particularly regarding whether it adopts an interventionist or more *laissez-faire* approach, and the balance it adopts between economic and social policy objectives – to create a range of possible situations (Lobet-Maris 1997, 1999). Not all of the possible outcomes that could be generated by such a schema were identified in our sample of countries. We could distinguish four particular situations amongst the nations studied: first are countries where there were dominant national champions (e.g. Deutsche Telekom and Siemens in Germany) which were at the centre of discussions of public policy. In the case of Germany this was coupled with an interventionist model to create what could be described as a corporatist framework; second, we find countries in where there is no clear pattern of dominance amongst the large and more pluralistic array of technology and media companies. In the case of the UK this is coupled with a laissez-faire policy tradition, to create what could be described as a highly competitive framework for ICT innovation. In the case of Japan an array of competing corporations (more oligopolistic rather than pluralist) interact with a state that is highly interventionist in technology development strategies in what could be described as a collective development model; finally, we find countries like Belgium, in which the weak development of local industrial players means that a state with an interventionist tradition may be drawn into a wider role as a 'proxy' for the private sector (Lobet-Maris 1999).

HOW THE NATIONAL CONTEXT SHAPED UNDERSTANDINGS OF ICT

The national studies had a number of particular concerns and foci. At the most general level the studies were concerned to look at general concepts and understandings of ICT that were emerging – and in particular to establish whether there was a *dominant image of ICT* within a country (or whether a number of competing conceptions prevailed), and *what this was* in terms of the concepts of how the technology infrastructure was to be configured, the kinds of application that would run upon them; the conception of who would be the users and how the technology would be used – and thus its social implications.

When this study was being launched, a point of sharp debate in this area concerned the metaphor and organising principles that were to be applied in relation to the delivery system/platform for mass-market ICT applications – particular into the home – where a dichotomy had been suggested between views of ICT as an extension of the World Wide Web (i.e. based on a networked, Internet-capable personal computer) or whether it was seen as an extension of the set-top box for digital TV.

These contrasting technical models – World Wide Web versus interactive TV – were associated, more or less implicitly, with contrasting visions (or more precisely suites of visions) of the use and the user of ICT. These different sets of views were associated with the influence and involvement of different sectoral actors; with different concepts of the uses and utility of ICT; and, with sharply contrasting moral visions of the information society. Thus the set-top box, with its limited technical interactivity (e.g. if operated by a TV remote control device rather than a keyboard and with its limited 'back channel') seemed to place the user in a passive consumer role – ordering pizza from your armchair – in contrast to the socially valorised activities of communication and information search through the personal computer linked up to the Internet.

Today this dichotomy seems less clear than when the SLIM research project was being developed. Fewer people believe that a single platform will emerge for ICT in the house. Instead, we will continue to have a range of different devices for particular purposes (as we already do, for example, with consumer electronic, where TV, radio and personal stereo continue as distinct products despite the technical possibility of integrating them. At the same time, the dichotomy between these routes has become blurred as key players in each market have sought to open up their offerings to the alternative option. Thus, for example, interactive TV is being proposed as a point of access, rather than an alternative, to the Web (Stewart 1999a). Finally, the growth of mobile telephony and other wireless applications, has stimulated a shift in attention from the locales of the workplace and home to the multiple settings of everyday life and to mobile access.

In some countries, our surveys were able to identify relatively clear national images of ICT – and conceptions that could be related to the relative strengths of these apparently counterposed concepts. Notably, in Germany, digital TV was placed centre stage – a development that may be related to the way the German ICT landscape has been dominated by two large media conglomerates – the Kirch-Group and Bertelsmann – and Deutsche Telekom (Kubicek, Schmid and Beckert 1999). In Norway, in contrast, the 'Internet model' prevails in a context in which many people already have access to the

Web from their homes (Brosveet and Sørensen 2000). Norway is rather open to the technological visions emerging from the USA – and, apart from Telenor, there are not powerful national champions that could shape national perceptions of ICT around their particular offerings.

However, in most countries, a range of models of ICT and its uses seemed to co-exist, ranging from established CD-ROM technologies, to the emerging technologies of the Internet and the World Wide Web. Our studies yielded some rather interesting differences in emphasis of national images on these two technologies between, for example, Denmark, in which ICT has been primarily understood in terms of CD-ROM and a country like the Netherlands in which CD technology is today taken for granted.[12] In Ireland both CD ROM and the Internet figure, while in Belgium and France attention is mainly focused upon the Internet. In the UK and Netherlands we see markedly pluralistic conceptions of ICT. It could be misleading to seek to draw too much inference from these accounts of national image – which may be influenced by recent public relations activities and initiatives (for example, in Switzerland, the idea of ICT is associated with the recent high-profile opening of the new Olympic Museum). In most cases the prevailing conceptions of ICT are poorly bounded and defined, and rather heterogeneous.[13]

We would seek to draw rather different conclusions from these complex and confused national patterns in terms of the perception of ICT. First the predicted convergence of information and communication technologies has not really materialised. The World Wide Web is showing epidemic growth rates across all countries surveyed (though there are important differences on penetration and growth rates). However, this will not be the end of the process – as evinced by the emergence of mobile telephony and growing attention to the emerging generation of wireless-based services, The collapse of the dot.com bubble highlights continued uncertainty about the business models that will prevail.

The search to identify national models/images would imply some level of resolution of the lines of development of ICT technologies and uses around relatively clear models (which could be global models or specific national ideal-types). In some circumstances, of course, particular national players may be strong enough, or associations amongst players may have sufficient coherence and influence, to shape overarching views. Germany provides one such instance. France has provided another, for example in the case of Minitel, though there has been some rethinking of the desirability and feasibility of such strategic state-sponsored initiatives following the difficulties with the French Cable Plan (Lobet-Maris 1999). However, in most contexts ICT refers to an extremely diverse, dynamic and turbulent set of developments. It involves a wide range of players, involved in content provision and application domains as well as diverse technology supply. In such a situation it seems implausible that specific models will come to prevail.

This could, of course, be taken to reflect that fact that we are a transitional situation in a context in which 'technological closure' has not yet taken place

OBSERVATIONS AND CONCLUSIONS

A substantial set of empirical findings has been amassed through the national studies of ICT uptake. Though some of this has been published elsewhere (Slack and Williams 1999, Jaeger, Slack and Williams 2000) it is useful to state here (briefly) some immediate conclusions from this inquiry – in relation to the divergence or convergence of perspectives, and also in relation to the role of the state.

Divergence or convergence

In contradiction to the convergence thesis, we note first that there are some clear differences in the way that ICT has been conceived in different countries – both regarding the extent to which dominant conceptions of ICT products and services have emerged, and in the nature of those conceptions.

Second, these differences can be related to the diverse social, historical institutional and technological settings for the development and use of ICT within different nations and regions, and the arrays of players involved. There is not, however, a simple relationship between particular factors and actors and the pattern of ICT developments. This comparative review highlights the complexity of interactions, and the way that particular influences may be mediated by other historical factors. This could be illustrated by the way that the influence of the size of linguistic groups depended upon how this factor might be taken up within political concerns. In this content, it seems rather difficult – and indeed unhelpful – to attempt to produce a simple characterisation and explanation of national patterns of technological change in ICT.

These observations serve, however, to underline the shortcomings of accounts which see technology supply as a universalising force which cuts across and gradually homogenises different social settings. There were, of course, some largely common technological developments across countries – the core ICT technologies are after all largely available at a global level. Certain platforms have been extraordinarily successful, such as the Internet Protocol, the Windows computing environment and GSM mobile technologies. However, the anticipated convergence around global technological models has not taken place. Instead, ICT technologies emerge as complex assemblages of global and local elements emerging in a process in which generic ICT capabilities are appropriated and incorporated to particular purposes within particular social settings. One important 'differentiator' arises through path dependencies (primarily arising as a result of 'sunk investment' particularly by final consumers). This was illustrated in the case of France, in which the earlier success of Teletel/Minitel – a large installed base of terminals and an established information market – constitutes a key factor shaping future ICT developments, for example holding back the uptake of PCs and the Internet (van Bastelaer 1999). The different outcomes of Minitel in France and Germany (with the latter able to migrate more readily to the Internet as an unintended consequence of the choice of an industry standard PCs, rather than a dedicated terminal, as the domestic Minitel terminal [Schneider 2000]) highlights the potential influence of particular contingencies.

Third though there does seem to have been some evidence of rhetorical convergence at the level of public policy discourses, this has only been partial. Some elements of technology policy recur: conceptions of the capability of the new ICT systems that will emerge; the economic and social benefits that will accrue from their adoption; and the competitive dangers of being 'left out'. Common patterns can also be detected in the areas and activities in which public-funded digital initiatives and experiments have been set up (for example in education and public information) (Jaeger, Slack and Williams 2000). However, examination also reveals continued differences in historical national policy styles and traditions – notably regarding the role of the state, between interventionist and more laissez-faire models. There is some evidence that the broader preoccupations of particular nations may be taken up and projected within ICT and information society policies.

These points reinforce the relevance of the social shaping of technology perspective to understanding how ICT technologies may be appropriated differently in particular national and regional settings. We have tried to map out these factors here. More detailed comparative analysis will be needed to capture the dynamics of this process – and the way in which regional and national settings shape technological innovation and policy-making.

The national studies, presented in Williams and Slack (1999), provide an important resource for researchers and decision-makers. They draw attention to both differences and similarities in the perceptions of emerging ICT technologies by a range of actors in different national settings. What is most striking is the *diversity of these responses* to an ostensibly global technology, both in terms of images of ICT technology and approaches to ICT policy. Cross-national research has a particular value here, as it allows us to open up for critical examination precisely those things which are taken for granted within particular societies and cultures. By drawing attention to the range of possible approaches in public policies and technology development strategies for ICT we hope to contribute to more informed discussion and debate.

The role of the state

As our cases show, the state continues to be a key player in promoting and providing resources for digital experiments. On the other hand, digital experiments are typically collaborative projects, bringing together various players – and most digital projects are 'hybrids'. Public policy and policymakers are thus important features of the translation terrain, and there are important issues in managing the public–private boundary, regarding how to integrate public and private elements.

One striking feature of the SLIM cases, and more generally of the wider array of digital experiments that we covered in our national reviews (Jaeger, Slack and Williams, 2000), concerns the key role of the public sector (at international, national and local levels) in supporting digital experiments. In some areas, notably cultural content products with well-established commercial markets (games, television), private provision did prevail. However, even here the state was frequently an important player whether as a key source of funding (as in the case of the Nerve Centre) or a provider (for example the public broadcaster RTE with its Irish 'heritage' projects). In the other areas we examined – education and public information – it is perhaps less surprising that the state was a central player in many if not all cases.

This finding is somewhat paradoxical given the contemporary emphasis across the various national information society policies produced in the last decade in the European Union, the USA and Japan on the leading role of the private sector in many areas of ICT provision (Williams and Slack 1999). Alongside their laissez-faire emphasis, these policies discuss the need for state support in those cases where the self-interests of the players concerned will not guarantee development (i.e. where markets may fail). Candidates for public funding include groups which lack economic power to ensure their needs are met, support for the early stages of research and or development of a new technology where the costs and risks may exceed expected benefits and finally the many public services where the state is itself the customer. Our evidence suggests that public sector support continues to be important across many application areas (Jaeger, Slack and Williams, 2000).

However, the boundaries between public and private provision seemed to be shifting. Partly as a result, digital experiments were most often 'hybrid' initiatives, involving public and private players and resources.[14] An important aspect regarding the continuation and growth of digital experiments concerns the establishment of a sufficient market of users to make a project self-funding. Dealing with these commercial issues and negotiating the public–private boundary may be a key issue in terms of the viability and success of a new ICT product. This has important policy implications, to which we return in Chapter 9.

NOTES

1. This approach has been extended to the G7 countries – and globally, for example through the Global Bangemann Challenge.
2. Where possible, resort was made to international (and especially European) databases and statistics to provide comparators and indicators. However, our work with these shows that there may be continuing problems with the use of such data, particularly in such a rapidly changing field (where for example statistics on the use of the Internet need to show the *month* for which the data pertains rather than the year – so great is the rate of growth in some countries), and where definitions are not sufficiently well established to provide benchmarks for comparisons.
3. For example, the countries covered in the review of digital experiments (see below) differ greatly in their size and the number of initiatives going on. In the larger economies, it was necessary to have a more selective focus than in the smaller ones. Thus, the review of digital experiments for the larger countries encompassed what was judged to be the most significant and interesting national cases, together with an array of more modest cases in the immediate locality, whilst in the smaller economies it was feasible to attempt a more inclusive review.
4. Each SLIM research centre undertook a study of its own national context. In addition, France was studied by the Belgium SLIM Centre, and Japan was studied by the UK SLIM Centre.
5. Thus the national studies and the 'Integrated Studies' both addressed the national/local adoption of products developed elsewhere. In addition a limited amount of specific research was undertaken of international developments. Insights were drawn from a related study at the Technical University of Denmark of the role of different interests and players in the global development of ICT standards and their implications for European ICT industries. FUNDP – Namur University studied EC policies for ICT, and compared them with those developed in the USA. Edinburgh University conducted a review of Japanese approaches to ICT to establish its distinctive features and differences/commonalities with European approaches.
6. Although some large industrial ICT users (e.g. in the finance sector) did appear as key drivers in the national studies, this was not the case with 'final-users' consuming and using ICT in their working and everyday lives. As we see in the Integrated Studies, these actors, widely dispersed across different workplaces, households and communities, were of crucial importance to the uptake and evolution of ICT products and services. Despite this they were not directly represented in the national public debates and interactions shaping the uptake of ICT. This raises an interesting question about who represents the final consumer/user – particularly in a context that these technologies are promoted in terms of the substantial benefits they will bring to members of society.

7 Equally, policies might be geared to preventing technical barriers to free trade – for example by requiring the provision of gateways and interoperability between competing networks (see Spacek 1997).
8 Paradoxically France's long-standing policy of combating Anglophonic domination in cultural products has not been applied strongly in relation to ICT products and services – partly perhaps because, through her earlier Minitel initiative, France already has a well-established information market (van Bastelaer 1999).
9 The analysis of the French situation presented by van Bastelaar and Lobet-Maris in Williams and Slack (1999) preceded the actions and initiatives taken by the Jospin government.
10 Thus the Dutch 1994 National Action Plan states that the government '*does not intend to produce a blueprint for the exact design and scope of the information superhighway ... Only society, in all its diversity, can continually decide which way developments will go*' *Information Superhighway – From Metaphor to Action. Action programme*, Part I (1.2. The government's role), Dutch National Action Plan (1994), http://info.minez.nl/docs/nap-en.htm.
11 Indeed, one of the features of liberalisation of telecommunications and the privatisation of national telecommunications service monopoly providers has been a move from a mode of control based upon direct political/administrative control towards a model in which private firms are subject to often extremely elaborate regulatory requirements (Mansell 1993).
12 Perhaps because it was there that CD technology was developed by Philips jointly with Sony in Japan.
13 It is notable that ICT applications such as gaming, though representing a significant market share, are often not considered in the mainstream debate.
14 We should not overlook here the contribution of voluntary actors, such as teachers and citizens, as well as the public administrators with responsibilities in the area.

9. Supporting Social Learning: Implications for Policy and Practice

This chapter discusses the implications of our findings for public policy and for the strategies of practitioners involved in ICT application. It starts by reviewing the evolution of technology policy and then goes on to address the current policy challenges, conceived in terms of shifting away from a technology supply-driven perspective towards a view that pays attention to technology appropriation and to reconciling the partially competing demands of technology dynamism and entrenchment. In the third section we begin to address more specifically the issues surrounding the exploitation of digital experiments. By focusing upon the key general 'social learning' question in relation to policy and practice – of how the knowledge and experience acquired can be transferred more widely – our discussion suggests new approaches to considerations of transferability of knowledge and a view that takes into account the many broader, indirect and longer-term outcomes of an experiment. The next section reviews our central policy recommendations in terms of how public policy can best support social learning around new technologies like ICT, drawing upon our set of studies digital experiments and ICT application initiatives across a range of settings. Finally, by way of a postscript, we then make a number of concluding observations and reflections about the achievements of the SLIM project.

THE EVOLUTION OF TECHNOLOGY POLICY

The failure of the orthodox 'linear model' of technology policy

The success of science-intensive industries in the decades after the Second World War, and the enormous advances made possible by 'Big Science' during the war, provided the rationale for the adoption across most developed states of a technology policy framework that looked to the promotion of advances in core sciences and technology to drive technical progress and economic growth. This policy tradition can be seen as an extension of post-Enlightenment views which treated technical advance as largely synonymous with social progress. This tradition tended to see technological change as a driver of social change – and embodied a further 'technological determinist' presumption that new technologies required particular sorts of social arrangements. The social policy issue was restricted to a very narrow one of how quickly and smoothly a society could adjust to meet the exigencies of new technology. It was argued that the biggest risks would arise from a failure to ensure rapid adoption and widespread uptake of technology. The role of social science was limited in this view to monitoring and predicting future technological trajectories, and in this way assessing the future 'socio-economic impacts' of new technologies.

Over the last two decades, the marked shortcomings of this traditional perspective have been progressively revealed. Both of its key elements – the emphasis on technology push and its inadequate model of the socio-economic and other implications of technological change – have been called into question.

Failure of science/technology push policies

Traditional technology policies in the European Union and most OECD countries pursued economic growth and competitiveness through support for Research and Technical Development (RTD). They presumed a 'linear model' of innovation, involving a one-way flow of information, ideas and solutions from basic science, through applied RTD, to industrial production and the diffusion of finished artefacts through the market to consumers. The basic driver of technological innovation was seen as advances in underlying scientific and technological knowledge, arising perhaps through public funded research and development, and the resulting creation of new technological artefacts.

However, this 'science and technology push' model proved not to be very effective in delivering successful technological advances, let al.one the economic and social benefits that were expected to accrue. Attention was drawn to a growing number of instances in which technologies developed in the laboratory were not taken up on a commercial basis. In contrast to the confident promises of the engineers, it turned out that new technologies often failed and, even when implemented, did not yield the predicted improvements in wealth creation and the quality of life. Technological innovation proved to be a rather uncertain endeavour. As we see below, as well as the narrowly 'technical difficulties' encountered, it often proved rather difficult to fit new technological offerings to social need and to build new markets (Williams and Edge 1996, Tait and Williams 1999).

As a result, public technology policies based on such linear models are increasingly seen as unhelpful – not least because of their privileging of technological supply and their artificial division of innovation into separate phases. Though the linear model is firmly out of fashion today, we should note that linear presumptions about the process of innovation are remarkably resilient.[1]

Lack of attention to unanticipated and undesirable outcomes of technology

The traditional model, with its emphatically positive attitude towards new technologies, tended to downgrade consideration of 'socio-economic outcomes' of technological change. To the extent that negative outcomes were acknowledged, they were largely presumed to be modest, and were seen as a necessary, but acceptable corollary of the presumed benefits of technology. However, since the 1970s, there has been increasing recognition that technologies could often bring *unanticipated* and *undesired* consequences. The starting point for this was growing awareness of the environmental and other risks of new technological activities.

The linear model also separated consideration of technology promotion from consideration of its costs and how they might be addressed. This was a consequence of its presumption that the costs and benefits of a technology were somehow inherent in the technology. In contrast, a growing body of critical discourse suggested that the particular socio-economic outcomes of a technology, its costs and benefits, were not an inevitable, technical matter but arose from the ways in which a technology was designed and implemented. This broadened the technology policy agenda, beyond a narrow concern solely with technology promotion to consider also choices in the design and application of technologies. It raised questions regarding how to assess in advance the prospects and implication of new technologies and select appropriate technological options accordingly (Collingridge 1980, Mackenzie and Wajcman 1985, Williams and Edge 1996).

The search for an interactive technology policy framework

Coupling technology supply to markets ?

The need for more integrated approaches to technology policy (both public policy and commercial strategy) that link technology to markets was first signalled some 30 years ago by Chris Freeman. Freeman (1974) stressed the importance of 'coupling' between technology supply and its user markets – linkages which are today analysed by evolutionary economists as the basis for the learning economy.[2]

However, linking new technologies to the 'market' of potential users has posed considerable challenges, particularly in areas such as ICT in which an accelerating rate of change in core technologies has been associated with enormous turbulence in product markets. Questions regarding advances in core technical capacities are matched by questions about how these may be applied and used.

The search for an integrated technology policy framework

Technology policies are being reassessed across the developed world. For example, the European Community Fifth Framework Programme (EC5FP) represented a challenging rethink of previous technology policy frameworks. This signalled a break from the technology-push models of the past and instead proposed an integrated approach, encompassing technology promotion and exploitation, and geared towards broad social and policy goals of competitive and sustainable growth etc. The emphasis was upon interdisciplinary research, bringing together social sciences as well as natural science and engineering disciplines, in a targeted effort, guiding individual RTD and uptake projects and linking them into clusters around certain strategic technological targets (Bruce et al. 2004). Similar developments can be noted in the UK and other national policy frameworks.

The integrated model proposed by EC5FP builds upon and complements developments in thinking about the process of technological innovation – the critique of the linear model and the espousal of an interactive model of innovation, as well as its recognition of the need for interdisciplinary approaches which bring together insights from social and policy sciences as well as economics and the disciplines of science and engineering. However, the challenge posed by EC5FP has, arguably, run ahead of theoretical understanding and our ability to make generalisations about policy approaches and solutions.

Today we see developments on a range of fronts in terms of both technology policy frameworks and practices. A variety of accounts seek to move away from a linear understanding of the innovation process (see, for example, Caracostas and Muldur 1998). However, this endeavour throws up a range of difficulties. If the relationship between technology and society is not unidirectional and deterministic, but involves a variety of more or less loose linkages at different levels, the setting for technology policy becomes extremely complex. The scope of technology policy thus becomes broader and more diffuse; a wider range of players and issues are involved (particularly with a generic technology such as ICT that can be applied across a number of sectors and areas of human life). The boundaries between technology policy and other forms of policy become blurred. Thus an interactive model, which starts by highlighting linkages between technology supply and user markets, rapidly encounters questions regarding the different structures of markets, the diversity of intermediate and final 'users', and the negotiated and unstable character of demand as needs evolve in the face of social and technical change (Sørensen and Williams 2002).

A growing body of empirical evidence points to this complexity. Much of this work also highlights the consequent difficulties of effective policy intervention, deriving in part from the contingent nature of the outcomes of technological innovation processes. Innovation processes are various. For example, some writers have stressed the local and differentiated cultures of technology supply (Hård 1994, Vincenti 1994). Similar issues emerge about the local and diverse character of user communities and markets (let al.one the relationship between local developments and contexts of use and how they may be transformed into more global markets). Given the importance of local contexts and contingencies, analysts have moved away from seeking to make strong generalisations about the role of particular institutional and policy settings in supporting successful innovation; there has been a shift away from a search for 'cook book' recipes for success towards a processual approach (Clark and Staunton 1989, Williams and Edge 1996).

Processual approaches to technological innovation – social shaping and social learning

The 'social shaping of technology' perspective, and other processual approaches, would therefore be critical of over-generalised approaches to technology policy. SST accounts would tend to emphasise the need for detailed empirical investigation and would offer a critical interrogation of the validity of presumptions and generalisations drawn by extrapolation from other technological contexts. However, such approaches may seem to offer cold comfort for hard-pushed decision-makers in government and industry. The challenge that decision-makers would pose for SST surround its ability to go beyond exercising a solely cautionary and critical voice, and offer instead some stronger generalisations (Sørensen and Williams 2002). We see three sets of developments that offer some ways forwards in addressing this challenge.

First, is the growing body of empirical research that begins to offer a more adequate account of the process of technological innovation and how this may vary across a range of settings – in terms of differing technical domains; differing socio-economic contexts of use; and differing national, cultural, economic and policy contexts.

The diversity thus revealed belies any simple correlation between context, process and outcomes of innovation. This complexity at first sight seems to frustrate attempts at generalisation. However, these studies have also given rise to a rich conceptual framework which helps to map out points of regularity and similarity. SST has been extremely productive in its theoretical and conceptual contribution. We note some important attempts to make these insights useful to non-academic audiences. Various academic writers have sought to make their ideas available to broader audiences – for example through a 'tool-box' approach, whereby a range of frequently encountered processes and issues in the social shaping process are identified. These are offered not as 'one size fits all' generalisations, but as *simplifications* about innovation processes together with indications of the contexts in which they may be relevant – as tools for thinking and acting which can help guide decision-makers. See for example, Molina's 'socio-technical constituencies' model, which identifies issues surrounding the internal *alignment* of groups of diverse players involved in technology development (Molina 1989, 1992, 1994) or the 'strategic niche management' model advanced by Rip and co-workers at the University of Twente which explores how novel technologies may need to be protected to survive in a hostile selection environment/ technology regime (Rip 1995, Rip and Schot 1999).

Towards reflexive policy: from government to governance

Alongside this toolbox approach we can also note approaches targeted towards the reflexive activities of the players involved in the innovation process itself. This work focuses upon the adaptive behaviours and learning opportunities available to the practitioners involved. From a processual perspective, in a context that is complex and variable, an important question about the conduct and outcome of innovation process concerns the ability of individuals, groups or organisations (or, for that matter, an economy) to react creatively to particular exigencies and a changing context. These kind of considerations underpin the move in this research project from the social shaping of technology perspective to a 'social learning' perspective. This move also parallels a growing concern more generally with processes such as learning (Gibbons et al. 1994) and capability building (see, for example, Nonaka and Takeuchi, 1995), and with the growing salience of ideas of the 'knowledge economy' and the 'learning organisation'.

These developments go hand in hand with changing models of the policy process, away from the idea of government exercising direct control over innovation actors through its traditional policy formation and implementation mechanisms (e.g. by imposing rules and deploying resources) towards a model of 'governance' in which the emphasis is upon less formalised techniques – in which policy outcomes are co-produced through a close two-way interaction between regulator and regulated (through knowledge flows, the alignment of views) (Kooiman 1993, Pierre and Peters 2000). The governance model would seem better suited for policy in contexts like those found in relation to current innovation in ICTs that are characterised by complexity, uncertainty and contingency. We see the search for policy fora and instruments that are more flexible and adaptable, able to deal with the diversity of innovation settings. As we see below, the social learning perspective has relevance here in relation to the processes involved in the policy development and implementation process, focusing for example upon how lessons are learnt from particular experiences and how they may be applied more widely.

CHALLENGES ARISING FROM THE CURRENT CONTEXT AND DYNAMICS OF INNOVATION OF ICT

Shifting away from a technology supply perspective

The key policy challenge posed by the SLIM study surrounds the need, we identified, for a 'double shift' in the focus of technology promotion efforts, away from their rather exclusive concern with promoting the development and supply of core technologies, towards:

- the *appropriation* activities of intermediate and final-users; and, associated with this,
- the need to support the development (and appropriation) of *cultural and information content.*

The need for such a shift in focus has been taken on board to some extent within a number of recent public policy developments – including, notably, the European Commission Information Society Technologies (IST) programme.[3] This is an important and welcome shift. However, the IST programme still displays some features that seem to have been 'inherited', passed on from its roots in traditional EC technology-supply driven Research and Technological Development programmes, particularly in its emphasis on the creation of new technological artefacts as a means to solving a range of socio-economic problems.[4]

The SLIM project points, instead, to the need for public support to be geared towards ICT innofusion and appropriation as well as, and perhaps as an adjunct to RTD projects. Moreover, support for ICT uptake (which is currently largely cast in terms of 'awareness raising' – as if there existed a finished set of ICT capabilities, matching user requirements, which potential industrial or private users merely need to be made aware of) needs to recognise the experimental nature of appropriation. The implication would be that such initiatives should include provision for a creative effort around the innofusion and domestication of ICT (e.g. through digital experiments) and for the dissemination of such appropriation experiences to other appropriators and to future technology supply.

The outcomes of artefact-centred approaches – digital experiments and RTD projects alike – have often seemed rather disappointing. They have certainly not fulfilled the widespread expectations that such artefacts, once developed, would move rapidly towards roll-out (bringing economic and social benefits). This draws attention to the *shortcomings of the artefact design/development, centred model.* The low uptake in most cases (even in those cases which pay more attention to the user than conventional technology-driven approaches, such as social experiments and user-centred design, which seek to build more user knowledge into the designed artefact) can be linked to the linear model which still underpins them, with their linear model of the relationship between initial design and subsequent roll-out. This is not to deny that some products may (albeit infrequently) emerge which are adequately aligned with current and emerging user/market requirements, and which become commercially successful and socially significant. However, given the difficulties in prefiguring evolving needs, such alignment is to some extent serendipitous, whether it was arrived at largely by accident or arose through careful and informed design. This is perhaps a salutary reminder that we do not yet have a sufficiently reliably understanding, nor sufficient information at this stage, to determine unambiguously which would be the most effective technology-supply strategy between, on the one hand, user-centred design (building more knowledge of specific user contexts and purposes into design) and, on the other, a laissez-faire strategy based upon maximising the launch of new products onto the market for selective uptake and adoption by users. The optimum is, of course, likely to lay somewhere between these extremes – and will certainly be different for different kinds of technology application. Current orthodoxy in system design discussions favours the former. However, we should not overlook the potential of the laissez-faire strategy. Indeed, if we examine the Japanese experience, success in consumer electronics appears to rest precisely upon very high levels of incremental innovation within their domestic market from which a number of innovations have emerged that have achieved global significance – highlighting the importance of receptive and demanding markets as well as smart supply (Porter 1990).

The mainstream response, exemplified by the European Framework Programmes and Technology Foresight in the UK, has been to take on board the criticisms of supply-driven models by setting up a range of measures to support commercial exploitation of public-funded research and development. This has been described as a 'linear plus' model, as it seeks to redress the failings of linear models of innovation, for example through support for knowledge transfer from public sector research to industry, and thus in effect tries to get innovation processes to work in an essentially linear manner, driven by the supply of technological knowledge (Tait and Williams 1999).

Attempts to shift away from the linear, technology-driven model of innovation must engage with certain problems. In particular:

1. Roll-out and commercial launch of a new ICT product or service typically takes an order of magnitude more resources than initial technical development. However, resources for commercialisation have been less readily available than for technology development. (Similarly,

support for associated developments has often been overlooked – in the case of digital media this relates particularly to the development of cultural content and information services which is labour intensive and expensive, but which has not been well-supported).

2. The process of domestication/appropriation typically occurs over a much longer timescale than technology innovation. The core technologies underpinning the Internet and the World Wide Web have enormous technical dynamism, with annual new launches of new versions, and ever-shortening product cycles, to the extent that today the ICT-supply industry considers two to five-year-old offerings as obsolete. This stands in contrast to the rather different dynamics and timescales of appropriation processes. The appropriation of ICTs is becoming accelerated. Whilst the complete cycle of appropriation through which earlier ICTs such as television became widely adopted and incorporated in everyday life took a generation or more, more recent ICTs such as the CD have been more rapidly adopted, but still took well over a decade to establish their market (Winston 1998).[5] Uptake of mobile telephony (and particular applications, such as SMS) have been faster still. However, a sharp disjuncture still remains between the cycles of innofusion and the cycles of appropriation.

Different parts of the innovation system may thus be operating with different dynamics. The slow pace of appropriation processes have often been unhelpfully miscast, from the viewpoint of the rhetorics of technology supply, as 'a problem' – an obstacle of user resistance or non-acceptance which needs to be overcome – rather than as an essential and central part of social learning around a technology. We cannot eliminate appropriation by *fiat* (though we may be able to organise social learning more effectively); the selective local uptake and experimentation around use are vital to technology domestication. Conversely it would be extremely unhelpful to conclude that the rate of technological development of core components of the information infrastructure should be chained to the slow pace and rather conservative trajectories of appropriation processes for ICT products and services. Instead, innovation policy needs to find ways of addressing the different dimensions and time frames of social learning – between innofusion and appropriation. An *integrated* innovation strategy, of the sort suggested by linear plus views, does not necessarily imply a *monolithic* strategy. Policy may need to explore the different spaces in which social learning is taking place. Further, we may need to find ways to *uncouple* particular elements of technology development and appropriation, rather than try to link these mechanistically together (Tait and Williams 1999).

Reconciling technology development and appropriation; dynamism and stabilisation

This brings us on to consider the potentially competing requirements that may beset a technological system, and how they may be reconciled. This encompasses the tensions between stabilisation and entrenchment versus the dynamism of development of a technology and between the creation and appropriation of technological capabilities, and the tensions that may exist between different elements of a technological system (e.g. core component technologies; the technology platform/infrastructure; applications).

Let us start with the example of the development of a single artefact. Referring back to Figure 4.3, when we consider a particular stage in the development of an artefact we can identify two potential vectors for its further evolution, between:

- *technical enhancement* – efforts geared around technology supply (to deepen and refine the technological knowledges underpinning artefacts) and,
- *social appropriation* – efforts geared around the domestication, consumption and use of artefacts.

Figure 4.3 thus highlights something of the different directions in which the innovation of an artefact can be taken. These may be (to a greater or lesser extent) orthogonal, one to another. There may be a trade-off between expending effort on technical enhancement versus social appropriation. In Chapter 4 we reviewed a range of strategies. At one extreme, development of an artefact may be taken primarily towards stabilising technological development, building markets and promoting social appropriation, and at the other, consolidation of uptake and use may be set aside in favour of the further development of the state of the art of technology. However, our analysis pointed to the variety of possible technology 'development/ innofusion/domestication' paths, between these extremes, which involve different strategies for matching technical potentialities with user requirement to get a technology to work under specific user circumstances. This suggests that there may be a number of strategies to reconcile these different exigencies.

Tensions may thus exist between the innovation, innofusion and further evolution of technologies (for example the technology infrastructure/ delivery system) and the promotion of a stable and homogeneous market – particularly for information and cultural content. The dynamisms of technology development and innofusion processes will tend to lead to a patchwork of more or less unique configurations as a result of the historical accumulation of user acquisitions over time. This presents particular problems where an information infrastructure is emerging through innofusion. The Internet is precisely such an example of a large-scale technological system arising through a process that is widely distributed across many players (including the huge numbers of 'final-users' configuring their local ICT platform and providing 'content'). These processes have significant consequences. High levels of dynamism and turbulence around technological trajectories and standards will tend to encourage consumers to defer purchasing decisions. This may not only slow the rate of adoption but, more critically, may undermine the very viability of new services where these depend upon attracting a critical mass of customers (Williams 1995). This has important implications for the information/content market. It was precisely this kind of consideration that underpinned the decision by France Telecom to support and heavily subsidise the widespread adoption of its Teletel/Minitel videotext system, This included making terminals available to large numbers of domestic consumers at low cost, and creating a framework for service suppliers to operate (including provision for charging for services through the customer's phone bill). France Telecom succeeded in creating circumstances favourable to the diffusion of the technology, and to social learning in innovating new services. As a result a huge range of information services could be made available – many of which proved commercially viable (Schneider et al. 1991). In contrast, in other countries videotext dramatically failed to fulfil the predicted rates of uptake (Bruce 1988, Thomas and Miles 1990, Schneider 1991, Christoffersen and Bouwman 1992); where it prevailed, it remained restricted to smaller niches of business users (Schneider et al. 1991).

The innofusion-based model that we have identified does not necessarily provide unequivocal support for such concerted models of technology adoption. For example, as Schneider (2000) has subsequently noted, Minitel subsequently served as a barrier in France to the uptake of Internet standards. This perspective emphasises the highly complex, indeed chaotic, character of many contemporary innovation processes. Though allowing for *de facto*, industry standards may emerge, the innovation process often remains extremely dispersed and divergent, as different groups implement at different moments, technologies that are rapidly evolving in the course of a dynamic process of further technical innovation through highly dispersed and differentiated processes of adoption and appropriation of artefacts across a range of settings. The outcome of such innofusion processes in the technology platform will be an extremely heterogeneous technology infrastructure through the crystallisation of many specific local configurations. Such a socio-technical setting would constitute a very uneven and fragmented market for complementary products. We see this already in the problem presented to information service/application providers of supporting a plethora of standards and platform capabilities.[6] This also affects the usability of the system as well as posing particular problems for design.[7]

One commercial and public policy implication that follows from these considerations is that different components of the ICT system may be subject to different exigencies and requirements. The development of a mass information market may call for stabilisation and concertation of the technology infrastructure. Whilst the relatively small numbers of public and commercial infrastructure providers may be able to cater for the heterogeneity that will consequently exist across the network (through gateways and standards) this will be particularly an issue for the local technology platform, where the mass of end-users lack the technical and other resources needed to ensure they are aligned to standards. On the other hand, optimising technological innovation in the ICT platform may be best served by innofusion. Public policy needs to grapple with this rather contradictory terrain. This will not be easy – as we see below, there may be competing demands between different policy goals and between different technical fields.

In relation to technology supply, we have already noted in Chapter 1 the emergence of strategies, such as the development of 'architectural technologies' (Morris and Ferguson 1993) which offer longitudinal compatibility to reconcile the dynamism of technology supply with the stability of *de facto* standards. Though the classic examples of such standardisation and architectural technology have been in relation to core 'technical' components (e.g. microprocessors and hardware more generally), we have suggested that such factors will increasingly come into play in relation to the 'user surface' (the human–computer interface and 'the look and feel' of an application, the organisation of information and navigation metaphors and signposts) (Williams 1993). The segmentation of the technology infrastructure into different levels, the stabilisation and standardisation of the interface between them, and the emergence of interoperability standards more generally can all be seen as ways of allowing technology dynamism to take place within particular 'black boxes', without imposing unwanted change and new learning requirements upon other players (Fincham et al. 1994, Brady, Tierney and Williams 1992). This kind of 'black-boxing' effort can be seen as a strategy for economising upon social learning. And though our general argument points to the valuable contribution of social learning processes, we should not forget that innovation and social learning is expensive in terms of time and effort; it may be frustrating – indeed it arguably needs to be economised on wherever feasible!

In this connection a repeated finding from SLIM studies of the early Community Information Systems (based for example on text processing and bulletin board technologies) is that the changes in the human-computer interface, to take on board the availability of improved graphic capabilities and the design metaphors of the World Wide Web, caused a lot of disruption and uncertainty for users (notably DDS and Craigmillar Community Information System). This was particularly an issue for groups that were less technologically confident and committed. Change imposes new uncertainties and learning costs upon existing users who have invested in learning and becoming familiar with languages and procedures. Change in the look and feel of the user surface (for example about how to operate the user interface and how to navigate uncertain information terrains) could present a considerable obstacle to these groups – particularly those lacking confidence and with a narrow skills base. Of course, certain kinds of users – perhaps the more confident ones - may become bored with a system that seems dated – creating an incentive for providers to constantly redevelop their offerings. One idea that might follow on from this concerns whether it would be possible to provide 'continuity in change' – with (at least the option of) some stability in the look and feel of earlier versions in new launches of information systems and content product. As well as economising on learning around the innovation of a product, a further key challenge is to find ways to ensure that earlier social learning outcomes can be carried forwards when a new product or service is launched. This is often not achieved. Many 'new' technological systems may actually be less well developed then earlier, more mature applications. For example, there was a loss of ease of use in the switch from integrated bulletin boards to separate World Wide Web and email systems.

The relationship between local and global

We conclude this section by considering a related issue – regarding how the local and the global are handled in policy discourses and strategies. Technology innovation has traditionally been seen in global terms – and European technology policy has often been couched in terms of combating and competing with global competitors from the USA and Japan. In contrast, this study has demonstrated the enduring importance of the local – particularly in relation to certain kinds of cultural content concerned with local identities.[8] Rather than following widespread presumptions regarding the dominance of the global over the local, the SLIM study points to the interweaving of local and global elements – and the continual reworking of the relationship between them. In this context there are enormous opportunities for players at local, regional, national and European levels to contribute to the social learning processes that will enable us genuinely to domesticate global ICT capabilities and put them to the services of various social groups. Crude globalisation theories underestimate these opportunities.

However, the counter-position – emphasising local contexts and markets – does not provide a realistic alternative. The opportunities will, of course, be very different depending on the technology in question. Some component technologies (e.g. commodity computer and memory chips) are produced by a global oligopoly; others may have much more local markets. However, we saw that even products and services geared towards local constituencies turned out to have attraction for others further afield (the Irish diaspora seizing upon cultural content produced initially for people in Ireland; the outside users of Amsterdam Digital City), even in cases where these wider markets were neglected or underestimated. And networking events and media coverage, as well as supplier promotion activities, can lead to the rapid diffusion of concepts that become global poles of attraction, shaping visions, expectations and offerings. Careful analysis is needed of the interplay between the local and the global – for example to identify what local experiences and solutions might be applied more widely and how. Jørgensen and Sørensen (1999) have introduced the concept of arena to analyse the arrays of more or less local 'actor worlds' involved in particular technologies; how they are configured together; the strategies of the players to reposition themselves/reconfigure the arena and the consequent possibility of surprises – changes in local–global relations and patterns of influence in a technological domain. The commercial strategies of suppliers and public policies for technology support thus need to address the complex arenas in which ICT products and services are unfolding.

EXPLOITATION AND TRANSFERABILITY OF RESULTS OF DIGITAL EXPERIMENTS

Introduction

This section explores the issues surrounding the exploitation of digital experiments, and Research and Technology Development initiatives in the field of digital products and services more generally. It then examines the implications of our empirical findings, and the social learning perspective, for policies for promoting the development and uptake of ICT.

Transferability is a central concern in discussions of social learning. The concept of social learning seeks to analyse not only local learning experiences but the ways in which they may be communicated more widely and 'translated' to make them applicable in other contexts. Social learning, to be relevant, has to involve some 'change of state' or practice – to mobilise (or stabilise) a development. Particular experiences need to be carried forward and communicated more widely. This raises questions about how learning takes place; how the lessons that have been learnt and transferred; and, finally, about what structures and incentives exist that may enable this learning to go on?

These considerations have very immediate implications for public policy and commercial strategy – both in relation to the problems in exploiting digital experiments, and in the kinds of policy-setting that might be needed for effective experimentation and exploitation of experimental outcomes. Public and private funders of digital experiments have the commendable desire to ensure that the projects they are supporting do lead to useful outcomes, and can be more generally applied. The pursuit of assured delivery of demonstrable research outcomes seems, paradoxically, to have often had the consequence of promoting a search for concrete 'deliverables', and in particular channelling expectations towards the idea that the transferability of project outcomes will be through the wide dissemination and use of artefacts.[9] The focus on concrete deliverables has been *unhelpful*, we argue, for a number of reasons. First, it may discourage the kind of experimentation about *usage* that we have identified as key to the domestication of technologies. Second, the complexity and uncertainties surrounding commercialisation and appropriation of technology mean that this form of exploitation is often unsuccessful, at least in the short term. Third, and following on from this, production of a material artefact is but one of a number of potential outcomes from a research project (or even a commercial trial), and is not necessarily the most significant transferable element. Equally important, potentially, are *non-material outcomes*. These include the development of specific social relationships amongst diverse players involved in technology development and use, and knowledges of the processes of developing these relationships and technologies. Though the non-material outcomes of a digital experiment may be far more significant than the material artefacts generated, there may, however, be *particular difficulties in capturing and conveying them*. In particular, as we explore below, many of the lessons are extremely contingent – and therefore may be hard to communicate and generalise.

Some areas of national and European technology policy have begun to address these issues about the wider exploitation of RTD – notably through measures to support collaboration between ICT suppliers and commercial user organisations in the development of ICT applications, which reflects the growing recognition of the importance of such linkages in the 'learning economy'. However, the broader types of outcome have not always been fully recognised by technology policies and strategies. They also present particular difficulties for wider transferability and exploitation, for several reasons: many of the lessons appear highly contingent to local circumstances; many outcomes are tacit; the full relevance of outcomes may only become clear in the longer term. Given this element of unpredictability, the attribution of success of failure to projects may be a perilous task – as we see in the next section, many important lessons may arise from a project that ostensibly was a failure. As a result, public policy and RTD and exploitation strategies need to be designed around the social learning and innovation exigencies of particular application fields.

Artefacts – a promise of transferability

Many pilots and trials are geared towards the development of new ICT artefacts. In a number of cases this is because they were funded under RTD programmes (both national and international, where the European Community has had long-standing involvement), which have historically received substantial levels of state funding. In contrast, greater difficulties have been experienced in obtaining the resources needed for the development of *cultural products and digital media content* (let alone the resources needed for disseminating and building markets for such products), which are also expensive (Preston 1999), but where public support has been less well established.

However, the vision of developing a new ICT artefact holds out the prospect of the rapid wider dissemination of the product or service – offering an apparently clear and self-evident metaphor for the exploitation process, and one that contains within itself the promise of social benefits and economic gain. However, a brief examination of digital experiments – including those studied by SLIM and more generally – shows that such a linear exploitation trajectory is more the exception than the rule. For start, the costs of commercial development are enormous, even in those cases where technical innovation has been successful in creating a new artefact. The evidence from the commercial players is that the costs of roll-out and commercial launch of new, mass-market platforms are probably an order of magnitude higher than the technical development costs. Such investments are only likely to be supportable by the very largest global players.

One consequence of the prevalence of the linear model, and the focusing of public support upon the *development* of artefacts, has been the relatively limited availability of funding for exploitation.[10] RTD support tends to be finite – and there is a danger that, once a project comes to the end of the funding 'conveyor belt', there may not be resources to carry forward the development and commercialisation of the artefacts (and, no less importantly, to maintain the expertise and relationships developed amongst the collaborators). The assumption that private finance will become available to bring products so developed to the marketplace is very often incorrect; public research funding will typically only cover the earliest stages of development;

by the time this ceases, often the business case is still not clear and much initial development work still needs to be done. The danger, of course, is that developers may be drawn, by the ready availability of public funds to support RTD, into the further technical development of an artefact, or a wholly new development project, simply by the need to find follow-on funding – and thus be diverted away from the commercialisation and societal appropriation of existing offerings. Other modes of support (with larger levels and longer timescales of support) may be needed for the commercial launch and wider uptake of ICT artefacts – especially with respect to services and digital media content. This has further implications for the way in which public support is provided – for example, as we discuss below, RTD projects need to have more elaborate 'exit strategies' for when public support ceases.

The focus on artefacts has had a number of other unhelpful consequences from the point of view of social learning around ICT products and services in terms of the way in which digital experiments are conducted. There seems to be a general tendency for funders to seek to fund novel elements – and a tendency to see such novelty primarily in terms of technical novelty. Partly in consequence many ICT projects seek to utilise or carry forward 'state of the art' technology. This may, paradoxically, impede in the short term the prospects for the appropriation and market uptake of ICT products and services, to the extent that what is developed that may be out of alignment with current market opportunities which are rooted in the skills, the expectations and the equipment possessed by potential users. Equally, we have found some evidence that projects may be subject to limited 'degrees of freedom', such that an emphasis on technical development may divert energies and resources away from technology domestication. The conduct of digital experiments with a strong technology development focus was more likely to be resemble the mode of control than the mode of experimentation. Our research suggests that if the goal is to promote the economic and social benefits of wider uptake and use of ICT applications, a rather more effective starting point for public policy might be with strengthening the local ICT appropriation capabilities, and promoting innofusion rather than with technology development per se.

Though in this account we have stressed the factors that tend to undermine the prospects of linear, research-driven processes, we should not downplay the possibility of the exploitation of such artefacts, but instead should show how these emerge from complex interactive social learning processes – and, notably, of innofusion. For example, an important aspect of the astonishing technological dynamism in relation to ICT development has been the way in which artefacts can be unbundled from particular applications and contexts of adoption and relocated into a different socio-technical context. In the recent history of ICTs, we can point to the **rapid uptake of certain generic models** regarding communication services (email, the Web) and concepts of the human computer interface (windows, buttons, hypertext navigation models). What seems to be important in this process is the 'unbundling' of components with a generic applicability from their particular context of development and use (as exemplified by the way in which the World Wide Web shifts from being a vehicle for scientific interaction, to a tool for more general information search and exchange). Elements are 'decontextualised'. Specific features of applications and tools, relating to particular contexts and contingencies of adoption, are removed or made removable or reconfigurable. These components can them more readily be incorporated as elements of new technological configurations.

Often what is taken up is a concept of use, rather than a technical component – for example a metaphor, such as that of the digital city. Again a job of reinvention is needed to make it more generically applicable. As we see in the next section, this process of unbundling different learnt elements is critical to understanding how social learning takes place (especially regarding the transferability of lessons from digital experiments).

Non-material outcomes

Some of the most valuable outcomes of a digital experiment or trial are non-material, including, notably, its contribution to the development of specific social relationships and the acquisition of new areas and combinations of knowledge. The circumstances for developing and exploiting these may be very different than those for material artefacts.

Relationships

The development and uptake of new ICT products and services typically relies on the activities of a range of different players (suppliers of complementary products – e.g. technical components; content). This was rather evident in the studies of cultural content development in which different organisational and industrial cultures of ICT and of media/cultural activities were brought together (Preston 1999, Nicoll 2000). An important element is the development of knowledge of other relevant players – and the building up of mutual understanding, and possibly commitments and alliances. Further exploitation of an ICT development project may depend upon the continuation of these relationships – perhaps through a follow-on project – directed towards further 'technical' development or as part of an effort directed towards the dissemination and exploitation of a developed product.

The challenges of effective exploitation may be very different from those surrounding development. There may be a need to *transform existing relationships*. In particular, studies of ICT and other 'network technologies' involving complementary products is that the benefits of cooperation in the development phase (reduced uncertainties; shared costs) may be outweighed by the growing pressures for competition in the exploitation phase. The shifting balance between cooperative and competitive incentives represents just one way in which relationships may need to be transformed.

A critical feature in relation to social learning is that direct *relationships with (end-) users are unlikely to be transferable* to other settings (either in commercial exploitation or further development). One reason for this is that user relationships – particularly for 'end-users' – are likely to be localised (users may be drawn to take part in projects in their immediate geographical or institutional vicinity), and partly because their interest may be pragmatic, ad hoc and ephemeral. Consumers drawn to become involved in particular products and services may do this as a 'one off' in response to particular needs and concerns; they want to solve their own problems and may not be interested in providing generic solutions; they are not necessarily interested in technology per se, and thus may not wish to extend their participation with suppliers to series of other ICT projects. These problems of end-user relationships may apply less acutely to business users of ICT products and services – particularly larger firms whose interest in exploiting emerging new technological opportunities may motivate them to become involved. Smaller firms have experienced difficulties in this respect, as they may lack the financial and technical resources to engage the supplier community. They may interact only infrequently with technology supply, around their own episodic investment/innovation cycles. The failure of the 'learning economy' in this area – evidenced by the weaknesses of supplier–user links and engagement – has been addressed by public policy initiatives specifically directed to promoting such linkages and innofusion (e.g. in the UK and in Sweden), for example in relation to industrial ICT applications or product development. Such initiatives might usefully be extended in relation to digital media content and service provision.

Processual knowledge

The acquisition, sharing and deployment of various different kinds of knowledge is crucial to these processes of building and maintaining relationships and collaboration around the development, innofusion and appropriation of ICT. A key motive for collaboration is the need for groups with different specialisms and expertise to share knowledge in specific areas needed to solve problems: for example about the responses of users to particular offerings and about the capabilities, culture and offerings of other providers. This is well illustrated by the study of Compuflex, which highlights the importance to this global player of having a 'local' partner who understood local media consumption patterns and the differences between markets. In addition such collaborations may offer more generic knowledge about the processes involved in developing and exploiting ICT products and services, building socio-technical constituencies etc.

There are significant problems about the transferability of both these kinds of knowledge:

1. Some of the knowledge may be highly contingent to particular circumstances and stages of innovation; questions arise about whether and how it may be possible to extrapolate from this knowledge and apply it in other similar circumstances and more generally. In other words, how relevant and reliable would this knowledge be in other settings.
2. Some of the knowledge may be tacit and difficult to formalise and communicate more broadly.

The need for a managed approach to knowledge transfer

Given the tacit nature and local 'stickiness' (von Hippel 1988) of such knowledge and experience, it may be difficult to capture and broadly disseminate such knowledge in a way that is useful and applicable. Conventional approaches to the dissemination of formal (e.g. 'technical') knowledge may not be effective. To give an example, building lists or databases of digital experiments – though potentially a useful resource in identifying other players in a field and thus facilitating contacts and networking – are unlikely to provide the basis for conveying the complex understandings generated in experiments. Even simple case descriptions, which will tend to be 'heroic' accounts of success stories, are unlikely to convey what was learnt from an experiment. In relation to such complex conceptual deliverables, *a more active strategy is needed both for building the knowledge base and managing knowledge transfer*. 'Observatories' and interest groups may provide a useful vehicle for this kind of managed dissemination process (Lobet-Maris and van Bastelaar 1999).[11] These could provide the basis for the more critical and analytic approach which is needed for digesting and communicating ICT project experiences and their outcomes. For example, given the importance of the local context, there is a need to address the contingencies surrounding the conduct of projects, linking these, where possible, to particular features of the case (e.g. regarding the particular actors involved or their local context) and the difficulties and weaknesses encountered as well as strengths of the case. Equally important would be meta-level analyses of generic, and potentially generalisable, aspects of the conduct and outcomes of experiments. Such analyses should consider the influence of specific historical contingencies, and the consequent limitations and uncertainties surrounding the application of these findings more broadly. Similarly, as we see below, in discussing success and failure, particular care must be exercised in using concepts such as 'success' or 'best practice', since these are always judged against particular frames (rooted, for example, in earlier technological settings) which may be contested and displaced. The possibility that existing criteria and world views may be challenged in turn calls for a certain level of humility and caution in making policy pronouncements (for example the criteria chosen for addressing such features as success and failure, and their presuppositions, must always be defined as precisely as possible). It further suggests a shift in the way we gather information to support decision-making. Instead of collecting information to reinforce particular top-down prescriptions, we may need to allow for a more reflexive process of gathering rich descriptions of experience and context that can be a resource for actors in creatively addressing their particular problem settings. Such an approach would fit in with the broader shifts in conceptions of the policy process signalled by the concept of governance in which policy formation and implementation are co-produced (for example through public support for technology clubs and networks).

Given the difficulties in formalising and communicating social learning experiences, an important avenue for exploitation may take the form of 'embodied knowledge'. An important prerequisite for the exploitation of experiences may be the existence of enduring actors with the interest and capacity needed to exploit these experiences and apply these more generally. This reinforces our argument about the key role of intermediaries in this social learning, in conveying and applying experiences and lessons more broadly, and brings us back to some of the questions raised in Chapter 4 about which actors can take on board the role of bringing out these lessons (transforming this knowledge; extracting from complex experiences particular simplifications which can have wider applicability).

Another reason why a more managed approach is needed for knowledge transfer rests with the character of user communities. We have already noted that their interest in particular innovations tends to be pragmatic and shortterm, arising from particular contingencies and perceived needs. Particular user organisations of ICT applications will invest only infrequently in new systems (Webster and Williams 1993), and thus have limited and highly particular engagements with the technology supply-side. One policy which may help enhance the 'learning economy' is to encourage user organisations and other stakeholder to participate in networks or fora to share experience and develop expertise with which to deal with the next generation of technology. (Of course, many suppliers already support 'user groups'.

However, these are limited to particular suppliers and even products and their operation may be constrained by the supplier.)

Success and failure of a digital experiment

One of the implications of the social learning perspective is that attribution of 'success and failure' may be rather perilous affairs – not least because different parties are likely to have different commitments to and perspectives of a project, and may draw different kinds of benefits and lessons. Attribution of success or failure may thus be a relative question – and is often contested (as we saw in the case of the Craigmillar Community Information Service). Indeed, whether a project is formally defined as a success or a failure may tell you very little about the outcomes of the project, but may primarily reflect local micro-political contingencies, regarding which groups have control over the retrospective attribution of success to a project. Our study of educational technology projects highlighted the way that dominant actors were able to impose their criteria and notions of success and failure. Success depends upon the criteria applied. However, the study noted 'the focus within most projects we studied is on success rather than failure: failure is considered not to contribute to lessons to be learnt, but a waste of time, resources and energy.' (van Lieshout, Egyedi and Bijker 2001: 310). *It is only under very particular circumstances that sponsoring groups find it in their interests to define a project as a failure.*

The search for 'success factors' and to identify 'best practice' can therefore be misleading, as well as potentially uninformative. Success is determined against the particular sets of criteria and frameworks, often based upon our knowledge, which may be more or less adequate, of established technology products and services and of the goals and presumed effectiveness of existing policy regimes. It will be clear from the preceding sections that questions inevitably arise about how far we can reliably extrapolate from the templates and exemplars on which these criteria are based. Are these metaphors reliable in different settings?[12] Have the rules of the game changed over time? The conclusion of our review of 'digital city' initiatives was that it was not possible to identify a simple account of 'best practice' regarding the model of development, the reorganisation of the administration nor the management of the technical changes. The factors that seemed to underpin particular successes were often highly contingent to the particular circumstances of the case in a way which impeded extrapolation to other settings. The exploitation of experiences arising from experiments was thus rather complex – and knowledge about change *processes* seemed to provide a more reliable basis for transferability than correlations between specific factors and outcomes (Lobet-Maris and van Bastelaar 1999).

The outcomes of ICT innovations are often rather mixed. For example, the Videotron interactive TV system has closed – but the experience gained is seen as valuable (Curry 2000). The assessment of success may thus be rather complex. However, from the point of view of social learning, what is more important is not whether a project is assessed as successful, but whether it provides for flows of information and alignments of players. From this perspective, it is not necessary for a digital experiment to proceed to roll-out/widespread use for it to make a helpful contribution to social learning. Some of the most important learning may take place in relation to projects that failed to meet the objectives set from the outset or that did not proceed to roll out.

It remains the case, nonetheless, that public support for ICT initiatives (as with many other areas) is structured around more straightforward (and indeed simplistic) ideas about project outcomes, reflected in the criteria of success and failure that are adopted. Public funding for ICT (and other) initiatives increasingly seems to take the form of competitive tendering for awards for discrete projects and initiatives (in contrast to earlier models of public service based upon direct, long-term funding of in-house developments/services). This mode of project-based funding has proved rather flexible, cost-effective and effective in orienting players towards ICT and in promoting new kinds of collaboration.[13] However, there may be pitfalls. For example, proposers face the temptation to oversell project achievements (and, linked to this, may seek to oversell the capabilities of technologies). Both funders and proposers may unwittingly contribute to a consensus about what can be achieved which is unrealistic and does not pay cognisance to the difficulties and uncertainties that may beset developments in this field. Competitive funding of this sort seems to be strongly geared towards achieving public success, and this will tend to be defined in terms of fulfilling the goals set at the outset. Those involved in experiments may feel under pressure selectively to present outcomes in a way that matches existing policy goals, and perhaps even to *conceal* certain problems which beset an experiment – and thus also conceal potentially valuable lessons concerning how these were addressed – in order to keep external funders confident and happy with the project. External sponsors are likely to look for clear indicators of success in the projects they have funded – which, as we have seen, tends to be in terms of concrete deliverables. Public (and thus politically directed) funding may thus have some unintended consequences that may impair social learning.[14]

A general consequence of social learning – and an important feature of digital experiments – is that it is not possible to fully prefigure the outcomes of a project at the outset. The conditions for promoting social learning may indeed be in conflict with the desire of project funders to seek to establish and guarantee project outcomes in advance as far as possible. We see this, for example, in the tendency of many ICT projects to be 'verification experiments' – geared towards confirming the value of ICT products in meeting preconceived success criteria – rather than more open 'diversification experiments'. This was particularly notable in the Educational technology projects we studied which were unduly oriented towards the development of ICT artefacts rather than what was judged to be the critical task of locally integrating ICT within particular educational settings and activities (van Lieshout, Egyedi and Bijker 2001).

Unpredicted and longer-term outcomes

Finally, the rules of the game in ICT innovation – and thus the criteria for judging an outcome successful – may be changing. The history of ICT shows that time and time again, we have based our criteria for judging new technologies upon experiences of the last generation of technology implemented. Somehow the historical lesson – that the eventual significance and uses of many ICTs were far removed from initial concepts – has been repeatedly suppressed by the takenfor granted 'self-evident' nature of current perceptions of a technology's utility (Norman 1998, Winston 1998, Williams 2000). As a result, what is today seen as a paragon of success may tomorrow be called into question. The case of Teletel/Minitel in France provides a perfect illustration. We have already discussed how this case came to be seen as a success story:, the substantial support by the French state and France Telecom for the widespread uptake of the technology platform, particularly through subsidised public acquisition of Minitel terminals, and the creation of a broader infrastructure for commercial information services have been widely acknowledged as a recipe both for overcoming critical mass constraints in building a new technology infrastructure and for developing a large and vibrant market for information services (Schneider et al. 1991). However, later work by the same researchers tells a rather different story. Schneider (2000) subsequently pointed out the difficulties that subsequently arose in France in adapting to the emergence and global dominance of Internet standards and web-based services. In contrast, in Germany, where by accident of history a standard personal computer was used as the domestic Minitel platform rather than a dedicated Minitel terminal adopted in France, it proved much easier to open up videotext to the burgeoning information services of the World Wide Web. No one recognised the significance of this choice between a dedicated and a generic solution; it only became critical in the light of the subsequent development and uptake of the Internet. Historical hindsight may thus call into question what had been seen as a successful strategy.

When we come to assess the implications for assessing the outcomes and significance of a digital experiment this has several implications. First, the outcomes of a digital experiment have to be examined in the light of their context. Second, the outcomes, and thus the ultimate significance, of an ICT application project are likely to be unpredictable and only fully revealed in the long term. Thus we saw that early projects in the field of community information systems such as the ideas of telecottages (allowing telecommuting and distribution of technical skills to the rural periphery, that had been developed in social experiments in Scandinavia in the 1980s) and the Manchester Host (Brosveet 1998) and Amsterdam Digital City (van Lieshout 2001) were widely discussed and taken up as best practice exemplars and templates for contemporary ICT application projects. Our examination of how these projects were used as best practice exemplars in various SLIM cases also highlighted difficulties and pitfalls in determining the appropriateness of such exemplars (as was illustrated by the Frihus 2000 project in Fredrikstad, Norway, in which the influence of the above was seen as unhelpful).

This brings us finally to the question of the verification and assessment of experimental outcomes. It is clear from the above that this is a complex process. We must bear in mind indirect and unanticipated outcomes which may result over long time frames and in other settings. Benefit can be drawn not only from experiments that succeeded in meeting their goal, but also from those that apparently failed. This brings us back to our finding that *the utility of a digital experiment cannot be determined solely in relation to how the experiment is conducted*, but has to be assessed against its broader context and history of socio-technical change, and the specific challenges addressed. Let us examine a case where an experiment yields new artefacts and practices. Other broader and more diffuse forms of social learning may take place – for example as the system is rolled out. The utility and value of these *direct outcomes* will ultimately be *verified* by whether they are carried forwards and built upon or alternatively rejected and refuted. In other words, the value of an experiment is related to the extent to which the intense, focused and *organised* forms of social learning it affords adequately emulate the more diffuse '*organic*' forms of social learning that may take place subsequently and across the full array of groups that may be involved.

HOW CAN PUBLIC POLICY SUPPORT SOCIAL LEARNING?

The social learning perspective poses a number of challenges regarding the policy process. On the one hand, it criticises work which attempts to draw mechanistic policy conclusions, for example, around the linear model of technological change (whether technology-push or its mirror-image market-pull models), and highlights the uncertainties that surround the development of policy. At the same time it opens up new policy opportunities. However, the critical scrutiny that the social learning perspective encourages also problematises the process of policy-making. In particular it encourages a reflexive turn, and one that shifts away from making particular substantive policy recommendations towards an approach which seeks to foster certain policy-making processes.

A key issue here concerns the extent to which public policy can support effective social learning (and in particular the knowledge flows needed both to transfer social learning achievements and to apply them more widely). This was illustrated by our study of ICTs in Norwegian education policy (Case Summary 9.1: Teaching Transformed) which concerns the extent that government provides for effective social learning. There were some signs of a shift from traditional top-down state policy development and implementation mechanisms, which did not cater for local learning, towards modes of governance in which the state acted as a knowledge broker, allowing local learning experiences (in this case around the integration of ICT into education) to be communicated and applied more broadly.

However, merely flagging the importance of social learning does not resolve the policy problem. Social learning processes are far from straightforward – they raise new sets of considerations. Although in theory all experiences may yield lessons that are in some sense useful, this does not imply that all social learning outcomes are equivalent. We have seen how the concern by project funders to be seen to have supported 'successful' projects might have negative consequences in terms of favouring less experimental approaches, as well, as the suppression or dispersal of potentially valuable experiences from projects which appeared to be 'failures'. This reminds us that social learning is a contradictory and 'lumpy' process, and points to the possibility of 'unhelpful learning outcomes'. As we saw in the preceding section, there is a danger of inappropriate *mimicry* of external examples rather than adaptation to local circumstances. This might arise, for example, where developers and promoters become too closely committed to particular examples or concepts of technology which may prove neither feasible or appropriate (Clark and Staunton 1988).

Case Summary 9.1
Teaching Transformed – ICTs in Norwegian education policy

Though government plays a central role in education provision and resourcing, the complex division of responsibility and control between central, regional, and local government, and considerable discretion by schools and even individual teachers makes it difficult to manage change. This case examines social learning around policy and practice for ICT and the Norwegian system of education, arguing that the 'learning economy' (with 'forward and backward linkages' and complex interactions between players may not operate in the way expected. Top-down policy interventions 'learning by regulation' needs to be integrated with local learning to be effective.

The Ministry of Education sees the ICT challenge mainly as an issue of access to machines and formal skills, with a rather conservative view of their potential for transforming and improving the quality of teaching. The Ministry monitors policy implementation by collecting mainly numerical indicators regarding the stock of computers in different schools, the number of training courses for teachers, and ICT programme expenditure, and the number of institutions with plans for the use of ICT in education. However, these are rather superficial indicators with little bearing on the pedagogical achievements made through ICT. This is evidenced by case studies which highlight very different educational outcomes in schools with very similar numbers of computers which show that the number of machines is less important than access to computers (e.g. one school had few computers but achieved ready access by transporting them on trolleys to where they were needed), some enthusiastic teachers with ideas and plans (where training teachers was crucial), and an active strategy to integrate ICT in teaching. The cases show that communication between teachers and management is essential. Where local social learning loops are lacking, ICT has to be domesticated individually again and again (Aune and Sørensen 2001).

Modes of public support

We have argued that the key to social learning is whether players find it in their self-interest to act as carriers of this knowledge – applying or disseminating it more broadly. But how is such self-interest recognised and sustained? What is the role of the state? Some interesting policy questions follow on from this, as we saw in Chapter 4. For example,:

- How does the **constitution** of an experiment contribute to the prospects for the subsequent wider application of experience obtained in the project?
- Does the **institutional setting** provide support and incentives for such longer knowledge transfer and application? In other words,
- Is it in the continued interests of constellations to stay together?
- Will key **intermediaries continue to exist** (or will bodies be dissolved and actors transferred to different settings when project funding ceases, preventing the further utilisation of the knowledge they had acquired)?

Of course, where experiments are privately funded, the question of whether a development is in the mutual self-interest of participants is, at least in principle, resolved from the outset. With state support, the allocation of resources becomes in principle an issue which must be addressed. We are interested in the consequences of state resourcing both in relation to the conduct of an experiment and its further exploitation. As we saw in Chapter 8, despite the laissez-faire rhetorics of public information society/information superhighway policies, the state seems to play a key role across a wide range of areas of ICT development, both in relation to public information and services and to areas of 'market failure'. The latter reminds us of high costs and deep uncertainties surrounding ICT developments and markets. Private players may not find it in their short-term interest to embark upon developments, particularly where there are uncertainties about whether a critical mass will be attained to make a project viable. It is in this context that the resources provided by the state seem to be crucial.[15] State resources – knowledge and coordination efforts as well as financial support – underpin the bringing into being of many of the digital experiments. Furthermore such public support often provided the main resources deployed by the intermediaries that were at the heart of ICT constellations, in building and maintaining experiments.

Unhelpful learning outcomes

The first question to be addressed concerns whether modes of public support help (or may inhibit) social learning.

We have already seen that public support may have 'unhelpful' learning outcomes (for example the 'suppression of public failure' which may impede 'learning from failure'). Similarly, the *survival contingencies* of particular projects that are dependent upon external income may have unintended and undesired consequences on the behaviour of particular actors or groups.

Learning to be fundable

For example, one challenge for many projects, especially those depending upon state subsidies, concerns the need to *learn to be fundable* (Slack, 2000b). To enlist the support of external funders it may be necessary to cast a project proposal in terms which match the presuppositions of potential funders. This may be reflected in the adoption of unrealistic or inappropriate goals in order to bring funders on board. Sometimes the result could be that sponsors' goals (driven perhaps by their particular visions of the benefits of ICT) took precedence over user needs.

Once a project has been established, how can it continue to secure the resources it needs? It proved relatively easy for many of the public-funded projects we studied to get money for new ICT project developments. They found it much harder to secure long-term funding, e.g. to establish an ongoing service or continued support for an existing project (particularly since an increasing proportion of central state funding is allocated, often very competitively, on a project by project basis, as part of an attempt to improve the effectiveness of state support). Selectors look for novelty and additionality from new services – which may force those already involved in particular experiments and services continually to recast their offerings according to the exigencies of funds being currently disbursed to obtain further funding. Where project actors have continually to reapply to do new and novel things to gain renewed funding, there are dangers of a *drift in goals* – where, instead of projects developing cumulative achievements, new goals were adopted that undercut fulfilment of goals from previous rounds of funding (and especially the exploitation of the knowledge etc. obtained therein) – and 'project fatigue' where those involved felt that they had lost their sense of direction and mission. For example, the Craigmillar Community Information System (UK) experienced difficulties in maintaining continuous funding from external sponsors – and was forced continually to recast itself throughout its history (for example its latest existence is as a 'Teleport' – a base for high-technology training and point of access for business facilities).

Learning to be viable

Conversely, as a project moves towards completion, there may be a need to *learn to be viable*. This has historically been a problem for many state-funded demonstrator projects (Jaeger, Slack and Williams, 2000). One criticism that can be made of guaranteed state funding for an experiment is that it may divert participants' attention from the need to assess and build the future market needed to ensure the wider uptake and perhaps commercial viability, of a new product and service. This had certainly been an important failing of certain historical social experiments (Cronberg et al. 1991).

Some of our cases achieved viability by building an effective market of 'users'. Amsterdam Digital City provides a shining exemplar, as it succeeded in taking on commercial and other income-generating services, weaning itself away from state funding as this waned, and becoming self-funding.[16] However, this remained something of an exception amongst the array of cases we studied. Other cases continued to be dependent upon state support – and there would be questions about their future viability were this to be withdrawn.

Modes of public support to promote social learning?

Social learning may be said to reach fruition when a technology becomes disseminated and embedded within particular social settings. However, current modes of public support for ICT may not cater well for such evolution. In particular, there is a lack of fit between the relatively short-term discrete funding models that are typical of much public support (under the project-based mode of support) and the longer-term requirements of digital experiments driven by the dynamics and timescales of social learning. As we saw, after the Nerve Centre had developed its Colm Cille ICT product it experienced problems in building a wider market – and identified the need to embark upon an active process of encouraging its appropriation. However, public support for this 'post-project' stage proved difficult to obtain. This illustrates a problem experienced by many digital experiments in obtaining the kinds of longer-term funding needed to support the exploitation of technologies they had developed and, in particular, support for the appropriation, entrenchment and commercialisation of their offerings. The prevalent model of finite, project-based funding may give little scope for a project to become established, bed down and provide the basis for more far-reaching societal learning process.

One implication might be that current regimes of project-based funding might need to be reformed specifically to cater for these long-term requirements – for example by encouraging longer-term multi-phase funding – either from the outset or through extension awards as a development project comes to an end. In many countries there are more or less covert modes of public support for commercialisation. However, these tend to be separate from research support (in part because of the constraints on government support imposed by competition law, constraints that do not apply to 'pre-competitive research'). The social learning perspective calls into question this traditional separation between (imputedly pre-competitive) research and development and a competitive commercialisation phase (which by implication does not involve research).

In shifting away from technology-supply driven policy frameworks the social learning perspective may also suggest the need to support the production of complementary products. The example of cultural products may be a special case here. Their high up-front development costs may create disincentives to producing content geared towards minority cultures and small markets, raising the risk of market failure. There could be a synergy here between policies geared towards technology appropriation and those concerned with the preservation of cultural diversity in the face of the globalisation of the media economy.

How best to integrate public and commercial funding

Alternatively, provision of long-term public funding is not without problems as it may divert the attention of participants away from the need to become viable by building a user base. There are dangers of a culture of dependency expressed in the frequent failure of initiatives to seriously address the market for products, a tendency towards an approach dominated by technology supply. The question arises of whether different funding strategies can be devised that balance between public and private support, and between long-term security and the necessity for short-term survival – to offset such potential problems and encourage more effective social learning.

Here we need to look at the integration of public and private support. As already noted, many ICT projects were hybrids, involving various private and public sector players. We have already highlighted the need for systems of resourcing to provide incentives for social learning, in terms of the exploitation of experiences gained within a project. We should consider here both 'direct' exploitation – for example in terms of the entrenchment/commercialisation of new ICT products and services – and indirect exploitation in further trials and experiments. These two may be in tension, with the former perhaps requiring that existing project constituencies remain broadly intact and grow. For the latter this is not necessary. Closure of a project could, of course, be a vehicle for dispersing staff and disseminating their embodied knowledge more broadly. However, for this to happen it is necessary that the players continue to be involved in the broad field, in roles where this experience can be applied.[17] Either way the key question is whether the intermediaries involved continue to exist albeit in a different setting, with the necessary incentives and context to apply their experiences more broadly.

Overall we conclude that further consideration is needed of modes of public support for ICT. A more balanced approach is needed. Three important shifts are needed:

1. away from the prevailing focus on technology development, towards its appropriation;
2. away from a preoccupation with hardware and technical platforms towards content and service development; and
3. away from the provision of technical infrastructure to include also the development and exchange of skills and experience.

To illustrate the latter, participants in Cable School complained that it was easy to get hold of computers (which were kindly donated by a neighbouring IT manufacturer), but almost impossible to obtain the training they believed would be needed to allow teachers to explore the educational potential of this technology (Slack, 2000a).[18]

REFLECTIONS ON THE STUDY

Having explored a mass of empirical findings and considered their implications, it is useful to conclude by looking back over the intellectual journey that the SLIM project has undergone and reflecting upon what has been achieved, and what still needs to be addressed.

At the outset of SLIM research, when the research questions and perspectives of this project were being elaborated, they were based upon prevailing knowledge and opinion then available about the problems in developing ICT products and services and how these might be addressed. These highlighted the problems in developing new kinds of ICT application. As a reaction against presumptions that supplier offerings provided finished solutions that addressed current and nascent user needs, a major current of thinking saw this problem in terms of acquiring a more adequate knowledge of the setting and purposes of different kinds of 'users', and thereby building a more adequate model of the user into the design of ICT offerings. With hindsight we can see that this was an idealised and simplified schema. We found, for example, that the presumption that designers possessed relatively clear representations of the user and the user's setting was in almost all cases incorrect; instead, designers were working with rather implicit and poorly specified models of the user based on incomplete and often unreliable sources of information. Our analysis further called into question the 'design fallacy' with its presumption that the best way to improve the design and usability of ICT artefacts was by enhancing inputs into prior design. We found instead that there were a variety of social learning strategies including laissez-faire and evolutionary models in which users contributed directly to system configuration in the innofusion/appropriation of artefacts rather than prior design. Our study suggests a rather different perspective to the established view of supplier ICT offerings as 'solutions' to social need. Instead, we suggest they are elements which 'users' selectively appropriate and configure with other artefacts, meanings and practices, around their evolving purposes. In this sense ICT offerings are always 'unfinished', to a greater or lesser degree.

In short, though the SLIM project showed that social learning is important – it revealed a rather different empirical reality from that presumed at the outset. For example, another initial presumption was that digital experiments would provide a context in which suppliers, users and other players could meet and negotiate their requirements. We found that it was necessary to take this point further, and see all ICT developments as being in some way experimental, involving various forms of collaboration within and between organisations. The diverse forms for ICT initiatives confirmed both our sense that social learning was important, also that there was considerable uncertainty about how such learning should be organised. In using the term 'social learning', we do not wish to convey an air of smooth and progressive accumulation and sharing of knowledge. On the contrary, we found that the institutional arrangements were in most cases fragmented and subject to deep fissures and rifts – suggesting a terminology of socio-technical constellations rather than the more orderly and focused image conveyed by traditionally adopted concepts such as socio-technical constituencies or systems. We further found that these constellations tended to be clustered around either technology development or, more frequently, the configuration and appropriation of technology – and there was a marked absence of constellations bringing together users and developers.

In this context we identified the importance of intermediaries as key players in social learning, linking together players from different sites and acting as conduits for the exchange of knowledge and other resources, which turned out to be crucial for innovation. We also found that these roles were often informal and unplanned, but emerged instead through experience. Partly as a result, the importance of intermediaries and their potential contribution was often overlooked. Indeed, certain informal intermediaries were not properly recognised in the digital experiments we examined (for example the contribution of teachers to the application of ICT in education). Public policy needs to pay more attention to intermediaries as a resource for innofusion and for the appropriation/domestication of ICT. Many contemporary ICT developments are configured from combinations of more or less standard technology components. Development of novel technical artefacts may be less important than this local appropriation effort (and one particular finding of the study concerns the importance of appropriation constellations which have received little attention hitherto). There is a consequent need to shift technology support efforts away from narrow concepts of technology development, to include a stronger appropriation effort, focused upon innofusion and domestication. The development of digital content is also resource intensive and requires support.

We should remember that the SLIM study was conceived in a context in which linear, supply-driven concepts of the innovation system were coming under criticism, and more integrated approaches were being sought which gave more attention to commercialisation efforts and building markets, as well as the development of technology artefacts. However, the implication of that emerging view was that policies could be developed for better supplier–user coupling that might secure a harmonious integration of technology development and diffusion. In contrast, this study has revealed the tensions and contradictions surrounding ICT development, for example between the different dynamics of development of component technologies and of the appropriation of applications. Whilst an integrated approach to innovation policy (and business strategy) is indeed needed, this does not imply the adoption of a monolithic model that tightly couples these together. Instead, loose coupling may be needed between some elements within an overall framework. For example, strategies may need to acknowledge the differing dynamics between the development of core technologies, delivery systems/platforms, applications and the information and cultural content they hold, and thus allow greater autonomy between policy interventions across these fields. ICT is a generic enabling technology capable of application to an extremely wide range of activities; an appropriation-focused policy goes well beyond the confines of IT supply and is implicated in many detailed policies for these specific domains.

The work presented here contains probably the most comprehensive and detailed empirical base of research into ICT design implementation and use conducted to date. Technologies and their contexts of application have, of course, moved on since this work was undertaken. Questions may be raised about the relevance of the findings presented here to these new settings. Certainly the rhetorical claims of technology supply – of novelty, of having overcome the shortcomings experienced with earlier rounds of technological change – may encourage scepticism towards, and rejection of, existing knowledge. However, this rejectionist view has extremely damaging consequences for analysis, for policy and for practice. It leaves researchers and practitioners without a proper empirical base, and dependent upon the shallow and largely self-serving case studies and accounts produced by commercial players. Moreover, as we have argued, studies of past rounds of innovation have much to tell us. Indeed, we can see the same kinds of problem arising repeatedly under different generations of technology – though they may be played out rather differently under somewhat different socio-technical circumstances. And whilst the social learning perspective mandates against attempts to make linear extrapolations from one socio-technical setting to another, it does provide a sophisticated set of tools and concepts for reasoning about how one setting may differ from another, and thus for assessing the pertinence of different kinds of knowledge.

Today the technologies of the Internet and the World Wide Web are gradually losing their novelty, their sense of new-fangledness, their strangeness; they are becoming normalised. But at the same time new technologies and concepts are being strongly promoted. New technology infrastructures are emerging (incorporating mobile telephony and grid technologies) and ideas of ambient and ubiquitous computing are being pursued through ideas of wearable computers, chips on everything, dust computers. These pose new challenges, not just for technology development, but also in terms of social learning – in terms of innofusion, domestication and regulation – in matching these generic capabilities to particular socio-technical settings of use. We hope this book will continue to serve a purpose by providing both a framework and substantive resources for this creative effort.

NOTES

1 Though the linear model is out of fashion today, linear presumptions about the process of innovation are remarkably resilient in public discourse and in policy presumptions. Their policy attraction may in part result from the promise they appear to hold out of a world in which the development, uptake and outcomes of new technologies can be planned and controlled together with rather straightforward mechanisms for intervening – through support for technological development – an illusion of control that seems to be enormously enticing to public and industrial decision-makers (Tait and Williams 1999).
2 Although focusing in this account upon the shortcomings of frameworks which emphasised 'science and technology push', we should also note their mirror image in those policy approaches which gave primary emphasis to the role of markets. Such approaches became prevalent in the 1980s, notably in the UK. Reacting against the failure of large-scale state sponsored technologies (for example, supersonic air-travel) it was suggested that the state was not very effective in 'picking winners'. Instead, it proposed that such choices should be left to the market. However, the evidence from this episode was that markets could fail in certain circumstances and that resort to short-term market exigency still left unanswered problems of technology policy. The implication is that neither a 'science push' nor a 'market pull' model are adequate. Instead, a realistic and useful policy framework needs to address the interaction between them.
3 Thus under EC the Fifth Framework Programme (1998–2002), IST included Key Actions in the areas of Systems and Services for the Citizen; Multimedia Content and Tools; and New Methods of work and e-commerce.
4 Similarly, though the Fifth Framework Programme overall gives welcome attention to the commercial and societal exploitation of RTD, the legacy of earlier supply-driven EC programmes live on in the very structure of the programme – for example, the separation between RTD and support measures. This separation, in IST and in 5FP as a whole, is extremely unhelpful. For example, the 'take-up' measures should be seen not just as a practical matter but as a resource for socio-economic research and even future socio-technical research and development. Thus, we would argue, studies of innofusion and the appropriation of earlier RTD projects could be no less important a source for innovation as further RTD.
5 Even this only tells part of a much more complex story. CD technology, having failed in its original market as a video recording medium, became the industry standard for audio recording, and also a medium for computer storage – an observation which reinforces our argument about the scope for transferring ICT components to new and different contexts (Winston 1998).

6 For example, digital cities and community information systems migrated up to web-based standards, but still faced problems as to which functions and which generations of browsers to support. As we saw in Chapter 7, UK Bank needed to make difficult decisions about the basic technology platform it would expect for customers using its Internet banking system.
7 For example, not all browsers currently support scroll bars. Should designers forego top-end functionality or conversely create products that are not geared to or usable by 'bottom-end users'? Can designers simultaneously design for different levels of user functionality? Conversely, users may find it frustrating if the functionality of their system does not match that of the content provider, and confusing if the look and feel of applications differs depending upon which technology platform/point of access is being used.
8 Perhaps the most striking example in the SLIM study is the proliferation of cultural products geared towards local Irish history and culture. However, this case immediately illustrates the ways that local and global relationships are being constantly transformed and reworked, since these products have attracted enormous interest from the Irish diaspora in the USA, Canada and Australia in particular, who have access to these products through the Internet.
9 Arguably, one reason for the continued attractiveness of the linear model of innovation, and the associated idea of technology exploitation through the dissemination of artefacts, despite the many critiques that have been advanced, is that these seem to hold out a simple and apparently compelling metaphor both for the ways in which technology development effort can be exploited (and likewise of how this exploitation process, and the cost-effectiveness of research investments, can be monitored). The search for evidence of such linear exploitation is however often fruitless (and can lead to false judgements about the lack of impact of research). (Bechhofer, Rayman-Bacchus and Williams 2001, Tait & Williams 1999).
10 Though there have been important exceptions including the European Commission TIDE (Telematics for the Integration of Disabled and Elderly people) programme, which had a broader approach. See: www.**cordis**.lu/ist/ka1/special_needs/ library_**tide**_communication.htm
11 One example of how this might proceed arises from the SLIM 'Integrated Study' of Public Administration/Digital Cities, which proposed a 'vade-mecum of digital cities', to the 4th European Digital Cities conference in Salzburg in October 1998. Drawing on the SLIM research findings a number of propositions have been advanced which aim to provide useful and practical advice to system developers, builders and future inhabitants of digital cities. A further stage was to validate these proposals more broadly for other digital cities. (*Rencontre réelle de villes virtuelles* [FUNDP-Namur-June 16, 1998], PAI Seminar on State, Citizens and Market [FUNDP-Namur-June 25, 1998], 4th European Digital Cities Conference [Salzburg, October 29–30, 1998]).
12 Thus, as we saw in Chapter 5, one of the problems that beset the Frihus 2000 feasibility study has been linked to the *inappropriate* resort to templates from other projects (including 1980s Scandinavian telecottage experiments as well as projects in New Brunswick, Canada and Salford, UK [Brosveet 1998]) that were very different in size and ambitions and whose success had been debatable. This highlights a more general problem that these kinds of metaphors and exemplars are, inevitably *stereotypical* and simplified accounts of developments 'elsewhere'. A crucial pitfall in the search for best practice exemplars is that assessments of the

success of a project may differ sharply – for example between those directly involved, those in the immediate proximity who have some dealing with it, and those at a greater remove, who may be forced to rely on public pronouncements and third-party assessments. In this process there is much scope for the circulation, and acceptance as factual of interpretations that others, more directly informed, might consider erroneous.

13 For example, European Commission and other public support has had considerable influence in promoting ICT initiatives in the education sector (Egyedi, Lieshout & Bijker 2001).

14 We are by no means arguing that private funding is immune from such problems. However, the external goals of the private sector are rather narrower and more pragmatic. It could further be expected that commercial players have an interest in seeking commercial advantage from particular knowledges and links obtained in a project, whatever its public outcome, and may be happy to accept and draw *private* advantage by learning from failure.

15 This was, of course, recognised at an early stage in discussions by the Bangemann report, which noted (High-Level Group on the Information Society 1994: 12) that 'Competition alone will not provide such a mass, or it will provide it too slowly'. This was why it was necessary for the European Union and the member states to stimulate and fund *social experiments*. since 'Initiatives taking the form of experimental applications are the most effective means of addressing the slow take-off of demand and supply' (High-Level Group on the Information Society 1994: 18).

16 Though as already noted, the community information role of DDS was ultimately overtaken by the proliferation of competing services and only its role as a commercial Internet service provider remains today.

17 For example, when Cable Co decided that it no longer needed a local capability in promoting interactive services, but would concentrate instead on its core TV service, the key player left the field altogether and the huge body of contacts and expertise he had developed was lost – to the company and to the wide array of collaborative projects in which he had become involved.

18 Important criticisms have been made of ICT in education programmes precisely because of this emphasis on hardware – and neglect of the huge costs of retraining existing teachers (Soete et al. 1997). Pleasingly, initiatives such as the UK's National Grid for Learning (Department for Education and Employment 1997) seem to have gone some way in taking these criticisms on board.

References

Akrich, M. (1992): 'The description of technological objects', in W. Bijker and J. Law, (eds) *Shaping Technology/Building Society*, Cambridge, MA: MIT Press, pp. 205–24.

Akrich, M. (1992a) 'Beyond social construction of technology: the shaping of people and things in the innovation process', in M. Dierkes, and U. Hoffmann, (eds), *New Technology and the Outset: Social Forces in the Shaping of Technological Innovations* Frankfurt and New York: Campus/Westview, pp. 173 –90.

Akrich, M. (1995) 'User representations: practices, methods and sociology', in A. Ripp, T. J. Misa and J. Schot (eds), *Managing Technology in Society: The Approach of Constructive Technology Assessment*, London and New York: Pinter, pp. 167–84.

Akrich, M. and Latour, B. (1992): 'A summary of a convenient vocabulary for the semiotics of human and nonhuman assemblies', in W. Bijker and J. Law, (eds), *Shaping Technology/Building Society*, Cambridge, MA: MIT Press, pp. 259–64.

Andersen, E. S. and B.-Å. Lundvall (1988) 'Small national systems of innovation facing technological revolutions: an analytical framework', in C. Freeman and B.-Å. Lundvall (eds), *Small Countries Facing the Technological Revolution*, London: Pinter, pp. 9–36.

Anderson, B. (1997) 'Work, ethnography and system design' (Technical Report EPC-1996-103), in A. Kent and J. G. Williams (eds), *The Encyclopaedia of Microcomputers*, vol. 20, New York: Marcel Dekker, pp. 159–83.

Arrow, K. (1962) 'The economic implications of learning by doing', *Review of Economic Studies*, 29, 155–173.

Arthur, W. B. (1989) 'Competing technologies, increasing returns and lock-in by historical events', *The Economics Journal*, 99 (March), pp. 116–31.

Aune, M. (1996) 'The computer in everyday life: patterns of domestication of a new technology', in M. Lie and K. H. Sørensen (eds), *Making technology our own? Domesticating Technology into Everyday Life*, Oslo: Scandinavian University Press, pp. 94–124.

Aune, M. and Sørensen, K. H. (2001) 'Teaching transformed? Social learning and multimedia in Norwegian education policy', in W. Bijker, T. Egyedi and M. van Lieshout (eds), *Social Learning Technologies: The Introduction of Multimedia in Education,* Aldershot: Ashgate.

van Bastelaer, B., with Lobet-Maris, C. (1999) 'Development of multimedia in France', in R. Williams and R. S. Slack (eds), *Europe Appropriates Multimedia*, Trondheim: Norwegian University of Science and Technology.

van Bastelaer, B., Lobet-Maris, C. and Pierson, J. with Burgelman J.-C., Punie, Y. and Neuckens, F. (1999) 'Development of multimedia in Belgium', in R. Williams and R. S. Slack (eds), *Europe Appropriates Multimedia*, Trondheim: Norwegian University of Science and Technology.

van Bastelaer, B., Henin, L., Lobet-Maris, C. and Eveno, E. (2000) *Villes virtuelles: entre Communauté et Cité*, Paris: L'Harmattan (also available in English at http://www.info.fundp.ac.be/~cita/publications/SLIM/).

Bechhofer, F., Rayman-Bacchus, L. and Williams, R. (2001) 'The dynamics of social science research exploitation', (36) pp. 124–55 van den Besselaar, P., Melis, I. and Beckers, D. (2000), 'Digital cities: organization, content and use', in *Digital Cities: Experiences, Technologies and Future Perspectives*, T. K. I. Ishida (ed), Dordrecht: Springer-Verlag, pp. 18–32.

Berg, A-J. (1994a) 'Technological flexibility: bringing gender into technology (or is it the other way around)?' in C. Cockburn, and R. Furst-Dilic (eds) (1994) *Bringing Technology Home: Gender and Technology in a Changing Europe*, Milton Keynes: Open University Press, pp. 94–110.

Berg, A.-J. (1994b) 'A gendered socio-technical construction: the smart house', in C. Cockburn, and R. Furst-Dilic (eds), (1994) (*Bringing Technology Home: Gender and Technology in a Changing Europe*, Milton Keynes: Open University Press, pp. 165–80.

Berg, A-J. (1996) *Digital Feminism*, Trondheim: Centre for Technology and Society, Norwegian University for Science and Technology.

Berg, A-J. and Aune, M. (eds) (1994) *Domestic Technology and Everyday Life – Mutual Shaping Processes*, COST A4 vol. 1, Social Sciences, European Commission Directorate-General Sciences Research and Development, Luxembourg: Office for Official Publications of the European Communities.

Benyon, D. and Imaz, M. (1999) 'Metaphors and models: conceptual foundations of representations in interactive systems development', *Human-Computer Interaction*, 14, 159–89.

Bessant, J. (1983) 'Management and manufacturing innovation: the case of information technology', in G. Winch (ed), *Information Teechnology in Manufacturing Process: Case Studies in Technological Change*, London: Rossendale.

Bijker, W. (1993) 'Do not despair: there is life after constructivism', *Science, Technology and Human Values*, 18 (4), pp. 113–38.

Bijker, W. (1994) 'Sociohistorical Technology Studies', in Jasanoff, Markle, Petersen and Pinch (eds), *Handbook of Science and Technology Studies*, Thousand Oaks, London and New Delhi: Sage Publications, pp. 229–56.

Bijker, W. and Law, J. (eds) (1992) *Shaping Technology/Building Society: Studies in Socio-Technical Change*, Cambridge, MA and London: MIT Press.

Bødker, S, and Greenbaum, J, M. (1992), *Design of information systems: things versus people*, Aarhus: Computer Science Department, Aarhus University.

Brady, T., Tierney, M. and Williams, R. (1992) 'The commodification of industry applications software', *Industrial and Corporate Change*, 1 (3), 489–514.

Brosveet, J. (1998) 'Frihus 2000: Public sector case study of a Norwegian it highway project', Center for Technology and Society, Norwegian University of Science and Technology, Trondheim, *STS Working Paper 1/98 (SLIM working paper No. 3)*.

Brosveet, J. and Sørensen, K. H. (2000) 'Fishing for fun and profit: national domestication of multimedia: the case of Norway', *The Information Society*, 16 (4), 263–76.

Brown, B., Green, N. and Harper, R. (eds) (2002) *Wireless World: Social, Cultural Interactional Issues in Mobile Communications and Computing*, London: Springer-Verlag.

Bruce, A., Lyall, C., Tait, J. and Williams, R. (2004) 'Interdisciplinary Integration in Europe: the case of the Fifth Framework Programme', *Futures*, 36, 457–70.

Bruce, M. (1988) 'Home interactive telematics – technology with a history', in F. van Rijn and R. Williams (eds), *'Concerning Home Telematics'*, Amsterdam: North-Holland, pp. 83–93.

Buser, M. and Rossel, P. (2001) Telepoly: the risk of creating high-end expectations, in W. Bijker, T. Egyedi, and M. von Lieshout (eds), *Social Learning Technologies: the Introduction of Multimedia in Education*, Aldershot: Ashgate.

Büscher, M., Gill, S., Mogensen, P., and Shapiro D. (2001) 'Landscapes of practice: bricolage as a method for situated design', *Computer Supported Cooperative Work*, 10 (1), 1–28.

Callon, M. (1987) 'Society in the making: The study of technology as a tool for sociological analysis', in W. Bijker, T. Hughes and T. Pinch (eds), *The Social Construction of Technological Systems*, Cambridge, MA: MIT Press, pp. 83–106.

Callon, M. (1991) 'Techno-economic networks and irreversibility', in J. Law, (eds), *A Sociology of Monsters. Essays on Power, Technology and Domination*, London: Routledge, pp. 57–102.

Callon, M. (1993) 'Variety and irreversibility in networks of technique conception and adoption', in D. Foray and C. Freeman (eds), *Technology and the Wealth of Nations: The Dynamics of Constructed Advantage*, London: Pinter, pp. 232–68.

Campbell-Kelly, M. (1989) *ICL: A Business and Technical History*, Oxford: Oxford University Press.

Caracostas, P., and Muldur, U. (1998) *Society, The Endless Frontier: A European Vision of Research and Innovation Policies for the 21st Century*, European Commission DGXII Science Research Development Studies EUR 17665, Luxembourg: Office for Official Publications of the European Communities.

Castells, M. (1996) *The Rise of The Network Society*, Oxford: Blackwell.

Cawson, A., Haddon, L. and Miles, I. (1995) *The Shape of Things to Consume: Delivering IT into the Home*, Aldershot: Avebury.

Christoffersen, M. and Bouwman, H. (eds) (1992) *Relaunching Videotex*, Dordrecht: Kluwer.

Ciborra, C. (ed) (1996) *Groupware and Teamwork: Invisible Aid or Technical Hindrance?* Chichester: John Wiley.

Clark, P. and Staunton, N. (1989) *Innovation in Technology and Organization*, London: Routledge.

Clausen, C. and Koch, C. (1999) 'The Role of Spaces and Occasions in the Transformation of Information Technologies', *Technology Analysis and Strategic Management*, 11 (3), 463–82

Clausen, C. and Williams, R. (eds) (1997) *The Social Shaping of Computer-Aided Production Management and Computer Integrated Manufacture*, vol. 5, proceedings of international conference, COST A4, ISBN 92 828 1569 2, European Commission DGXIII, Luxembourg.

Cockburn, C. (1993) 'Feminism/Constructivism in technology studies: notes on genealogy and recent developments', paper to workshop on European Theoretical Perspectives on New Technology: Feminism Constructivism and Utility Brunel University, September.

Cockburn, C. and Furst-Dilic, R. (eds) (1994) *Bringing Technology Home: Gender and Technology in a Changing Europe*, Milton Keynes: Open University Press.

Cohen, E. (1992) *Le Colbertisme "high tech". Économie des télécom et du grand projet*. Paris: Hachette.

Collingridge, D. (1980) *The Social Control of Technology*, London: Pinter.

Collingridge, D. (1992) *The Management of Scale: Big Organizations, Big Decisions, Big Mistakes*, London and New York: Routledge.

Collinson, S. (1993) 'Managing product innovation at Sony: the development of the Data Discman', *Technology Analysis and Strategic Management*, 5 (3), pp. 285–306.

Collinson, S., Fleck, J. Molina, A., Stewart, J., Tomes, N. and Williams, R. (1996) *Forecasting and Assessment of Multimedia in Europe 2010+*, Final Report to European Commission, Edinburgh: University of Edinburgh.

Collinson, S. and Molina, A. (1998) 'Reorganising for knowledge integration and constituency building: product development at Sony and Philips', in R. Coombs, K. Green, A. Richards, and V. Walsh (eds), *Technological Change and Organization*, Cheltenham: Edward Elgar, pp. 76–107.

Collinson, S. (1999) 'The Development of Multimedia in Japan' in R. Williams and R. S. Slack (eds), *Europe Appropriates Multimedia*, Trondheim: Norwegian University of Science and Technology.

Coombs, R., Saviotti, P. and Walsh, V. (1987) *Economics and Technological Change* Basingstoke: Macmillan.

Coopersmith, J. (1993) 'Facsimile's false starts', *IEEE Spectrum*, February, pp. 46–9.

Cowan, R. (1992) 'High technology and the economics of standardization', in M. Dierkes and U. Hoffmann (eds), *New Technology and the Outset: Social Forces in the Shaping of Technological Innovations*, Frankfurt and New York: Campus/Westview, pp. 279–300.

Cronberg, T. (1992) 'Technology in social sciences: the seamless theory', mimeo, Technical University of Denmark, Lyngby.

Cronberg, T., Duelund, P., Jensen, O. M. and Qvortrup, L. (eds) (1991) *Danish Experiments – Social Constructions of Technology*, Copenhagen: New Social Science Monographs.

Curry, A. (2000) 'Learning the lessons of videoway: the corporate economy of new media trials', *The Information Society*, 16 (4), pp. 311–8.

David, P. (1975) *Technical Choice, Innovation and Economic Growth: Essays on American and British Experience in the Nineteenth Century*, Cambridge: Cambridge University Press.

De Certeau, M. (1984) *The Practice of Everyday Life*, London: University of California Press.

Department for Education and Employment (1997) *Connecting the Learning Society: The National Grid for Learning*, London: UK Department for Education and Employment.

Dunlop, C. and Kling, R. (eds) (1991) *Computerization and controversy: value conflicts and social choices*, Boston, MA: Academic Press.

Dutton, W. H. (1995) 'Driving into the future of communications? Check the rear view mirror', in S. J. Emmott (ed) *Information Superhighways: Multimedia Users and Futures*, London: Academic Press, pp. 79–102.

Dutton, W. H. (ed) (1996) *Information and Communication Technologies: Vision and Realities*, Oxford: Oxford University Press.

Edge, D. (1988) 'The social shaping of technology', *Edinburgh PICT Working Paper* No. 1, Edinburgh University.

Egyedi, T. (2001) 'Diversified hypermedia use: an experiment with dis-closure', in M. van Lieshout, T. Egyedi and W. Bijker (eds), *Social Learning Technologies: The Introduction of Multimedia in Education*, Aldershot: Ashgate.

Ehn, P. (1988) *Work-Oriented Design of Computer Artifacts*, Stockholm: Arbetslivcentrum.

European Commission (1996) Preliminary guidelines for the Fifth Framework Programme of research and technological development activities. *Inventing Tomorrow: Europe's Research at the Service of its People'*. COM(96) 332 Brussels.

European Commission (1996a) *Europe at the Forefront of the Global Information Society: Rolling Action Plan*, communication from the European Commission,

Brussels, Commission of the European Communities 27 November. 1996 COM(96) 607.
Fincham, R., Fleck, J., Procter, R., Scarbrough, H., Tierney, M. and Williams, R. (1995) *Expertise and Innovation: Information Strategies in the Financial Services Sector*, Oxford: Oxford University Press/Clarendon.
Fischer, C. (1992) *America Calling. A Social History of the Telephone to 1940*, Berkeley, CA: University of California Press.
Fiske, J. (1989) *Understanding Popular Culture*, London: Routledge.
Fleck, J. (1988) 'Innofusion or diffusation? The nature of technological development in robotics', *Edinburgh PICT Working Paper*, No. 7, Edinburgh University.
Fleck, J. (1988a) 'The Development of information integration: beyond CIM?' *Edinburgh PICT Working Paper* No. 9, Edinburgh University. A digest of this paper, prepared for the Department of Trade and Industry, is available as 'Information-integration and industry', *PICT Policy Research Paper* No. 16, Economic and Social Research Council, Oxford, 1991.
Fleck, J., Webster, J. and Williams, R. (1990) 'The dynamics of IT implementation: a reassessment of paradigms and trajectories of development', *Futures*, 22, pp. 618–40.
Forester, T. (ed) (1985) *The Information Technology Revolution*, Oxford: Blackwell.
Forester, T. (ed) (1989) *Computers in the Human Context: Information Technology, Productivity and People*, Oxford: Blackwell.
Fransman, M. (1998) 'Convergence, the Internet and multimedia: implications for the evolution of industries and technologies', in *Proceedings of the International Telecommunications Society 12th Biennial Conference*, (ITS-98) ITS: Stockholm.
Freeman, C. (1974) *The Economics of Industrial Innovation*, Harmondsworth and Baltimore, MA: Penguin.
Freeman, C., Clarke, J. and Soete, L. (1982) *Unemployment and Technical Innovation: A Study of Long Waves in Economic Development*, London: Pinter.
Friedman, A., with Cornford, D. (1989) *Computer Systems Development: History Organisation and Implementation,* Chichester: John Wiley.
du Gay, P., Hall, S., Janes, L., Mackay, H. and Negus, K. (1997) *Doing Cultural Studies: The story of the Sony, Walkman*, London and New Delhi: Sage.
Gibbons, M., Gibbons, M., Limoges, C., Nowotny, H. Schwartzman, S., Scott, P. and Trow, M. (1994) *The New Production of Knowledge: The Dynamics of Science and Research in Contemporary*, London: Sage.
Gilfillan, S. (1970) *The sociology of invention*, Cambridge, MA: MIT Press.
Graham, I., Spinardi, G. and Williams, R. (1996) 'Diversity in the emergence of electronic commerce', *Journal of Information Technology*, 11 (1), pp. 161–72.
Green, E., Owen, J. and Pain, D. (eds) (1993) *Gendered by Design? Information Technology and Office Systems*, London and Washington DC: Taylor and Francis.
Haddon, L. (1992) 'Explaining ICT consumption: the case of the home computer', in R. Silverstone and E. Hirsch (eds), *Consuming Technologies: Media and Inormation in Domestic Spaces*, London: Routledge, pp. 82–96.
Hansen, F. (1998) '*Blackout*', mimeo, Lyngby, Department of Technology and Society, Danish Technical University.
Hansen, F. and Clausen, C. (1998) '*Teleteaching on Bornholm*', SLIM Education Stream Report, mimeo, Lyngby: Department of Technology and Society, Danish Technical University.
Hård, M. (1994) 'Technology as practice: local and global closure processes in diesel engine design', *Social Studies of Science*, 24 (3), pp. 549–85.
Hebdidge, D. (1979) *Subculture: The Meaning of Style*, London: Methuen.

Hepsø, V, (1997) 'The social construction and visualisation of a new Norwegian offshore installation', *Proceedings of the ECSCW97*, J. Hughes et al. (eds) Kluwer Academic, 109–24.

Herbold, R. (1995) 'Technologies as social experiments. The construction and implementation of a high-tech waste disposal site', in A. Rip, T. J. Misa and J. Schot (eds), *Managing Technology in Society: The Approach of Constructive Technology Assessment*, London and New York: Pinter, pp. 185–97.

High-Level Group on the Information Society (1994) (Bangemann Report) *Europe and the global information society: Recommendations to the European Council*, Brussels, 26 May.

Hirschman, A. (1970) *Exit, Voice, Loyalty*, Cambridge, MA: Harvard University Press.

Hirschman, E. (1980) 'Innovativeness, Novelty Seeking And Consumer Creativity', *Journal of Consumer Research*, 7 (December), pp. 283–95.

von Hippel, E. (1988) *The sources of innovation*, New York and Oxford: Oxford University Press.

von Hippel, E. (1994) '"Sticky information" and the locus of problem solving: implications for innovation', *Management Science*, 40 (4), 429–39.

Howells, J. and Hine, J. (eds) (1993) *Innovative Banking: Competition and the Management of a New Networks Technology*, London and New York: Routledge.

Hughes, T. (1983) *Networks of Power*, Baltimore, MD, and London: Johns Hopkins University Press.

Hutchison, A. (1997) 'Empty icons in the metaphor trap', *ASCILITE '97 what Works and Why? Reflections on Learning with Technology*, Australian Society for Computers in Learning and Tertiary Education, 7–10 December, Perth, Western Australia, (sampled 15 September 2003) http://www.ascilite.org.au/conferences/perth97/papers/Hutchison/Hutchison.html

International Telecommunications Union (no date) *3G Mobile Licensing Policy: From GSM To IMT-2000 - A Comparative Analysis*, Telecommunication Case Studies, New Initiatives program, Office of the Secretary General of the International Telecommunication Union *c*. 2001, 25 September 2003. www.itu.int/osg/spu/ni/3G/casestudies/GSM-FINAL.doc

International Telecommunications Union (2002) 'Internet for a Mobile Generation', ITU Internet reports, Geneva: ITU.

Jæger, B. (1991) 'Digitale byer i København og Europa' (Digital cities in Copenhagen and Europe), in K. V. Andersen, et. al., *Informationsteknologi, organisation og forandring – den offentlige sektor under forvandling*, København: Jurist- og Økonomforbundets Forlag, pp. 89–110.

Jaeger, B. (2002) Innovations in public administration: between political reforms and user needs, in J. Sundbo and L. Fuglsang (eds), *Innovations as Strategic Reflexivity*, London: Routledge, pp. 233–54.

Jaeger, B. and Hansen, F. J. S. (1999) 'Multimedia in Denmark', in R. Williams and R. Slack (eds), *Europe Appropriates Multimedia*, Trondheim: Norwegian University of Science and Technology.

Jaeger, B. and Qvortrup, L. (1991) 'Community teleservice centres', in T. Cronberg, P. Duelund, O. M. Jensen and L. Qvortrup (eds), *Danish Experiments – Social Constructions of Technology*, Copenhagen: New Social Science Monographs, pp. 27–41.

Jaeger, B., Slack, R. and Williams, R. (2000) 'Europe experiments with multimedia: an overview of social experiments and trials', *The Information Society*, 16 (4), pp. 277–302.

Jakobs, K. and Williams, R. (1999) 'Proceedings of First IEEE Conference on Standardisation and Innovation in Information Technology (SIIT '99)' Aachen, Germany, 15–17 September, Aachen University, Aachen.

Jenkins, R. V. (1975) 'Technology and the market: George Eastman and the origins of mass amateur photography', *Technology and Culture*, 15, pp. 1–19.

Jørgensen, U. and Sørensen, Ø. (1999) 'Arenas of development – a space populated by actor-worlds, artefacts, and surprises', *Technology Analysis and Strategic Management*, 11 (3), pp. 409–29.

Kahin, B., and Keller, J. (eds) (1995) *Public Access to the Internet*, Cambridge, MA and London: MIT Press.

Kahin, B., and Wilson, E. (1996) (eds) *National Information Infrastructure Initiatives, Vision and Policy Design*. Cambridge, MA: MIT Press.

Kemp, R., Schot, J. and Hoogma, R. (1998) 'Regime shifts through processes of niche formation: the approach of strategic niche management', *Technology Analysis and Strategic Management*, 10 (2), pp. 175–95.

Kerr, A. (1999) *Social Learning via Trojan Horses, Social Learning in Multimedia, Cultural Content Integrated Study*, Dublin: Dublin City University (mimeo) http://www.vcss.ed.ac.uk/SLIM/private/I-studies/pp/pp.html

Kerr, A. (2000) 'Media diversity and cultural identities: the development of multimedia "content' in Ireland", *New Media and Society*, 2 (3), pp. 286–312.

Kerr, A (2002) 'The business of culture. New media industries in Ireland' in M. Peillon, and M. Corcoran (eds), *Ireland Unbound. A Turn of the Century Chronicle*, Institute of Public Administration. Dublin: Ireland.

Kerr, A. and Preston, P. (1999) 'The development of multimedia in Ireland', in R. Williams and R. S. Slack (eds), *Europe Appropriates Multimedia*, Trondheim: Norwegian University of Science and Technology.

Kooiman, J. (ed) (1993) *Modern Governance: New Government-Society Interactions*, London: Sage.

Kubicek, H., Dutton, W. and Williams, R. (eds) (1997), *The Social Shaping of Information Superhighways: European and American Roads to the Information Superhighway*, Frankfurt: Campus Verlag.

Kubicek, H., Schmid, U. and Beckert, B. (1999) 'The Development of Multimedia in Germany', in R. Williams and R. S. Slack (eds), *Europe Appropriates Multimedia*, Trondheim: Norwegian University of Science and Technology.

Kuhn, S. (1996) 'Educating software designers using a studio approach: an exploratory study' *Proceedings of the 4th Software Cultures Workshop*, Vienna, 7–9 November, Technical University of Vienna, ISSN 1021 7363, pp. 143–6.

Laegran, A. S. and Stewart, J. (2003), 'Nerdy, trendy or healthy? Configuring the Internet café', *New Media and Society*, 5 (3), pp. 357–78.

Latour, B. (1986) *Science in Action*, Milton Keynes: Open University Press.

Latour, B. (1988) 'How to write "the prince" for machines as well as machinations', in B. Elliott, (ed), *Technology and Social Process*, Edinburgh: Edinburgh University Press, pp. 20–43.

Law, J. and Callon, M. (1992) 'The life and death of an aircraft: a network analysis of technological change', in W. Bijker and J. Law (eds), *Shaping Technology/ Building Society: studies in socio-technical change* Cambridge, MA, and London: MIT Press, pp. 29–52.

van Lente, Harro (1999) 'Development of multimedia in the Netherlands', in R. Williams and R. S. Slack (eds), *Europe Appropriates Multimedia*, Trondheim: Norwegian University of Science and Technology.

Lie, M. and Sørensen, K. H. (eds) (1996) *Making Technology our Own? Domesticating Technology into Everyday Life*, Oslo: Scandinavian University Press.

van Lieshout, M. (1999) 'The digital city of Amsterdam: between public initiative and private enterprise', in C. Lobet-Maris, and B. van Bastelaer (eds), *Digital Cities Final Report*, mimeo – Faculte Universite de Notre Dame de la Paix, Namur: CITA, pp. 61–110.

van Lieshout, M. (2001) 'Configuring the digital city of Amsterdam', *New Media and Society*, 3 (1), pp. 27–52.

van Lieshout, M., Egyedi, T. and Bijker, W. E. (eds) (2001) *Social Learning Technologies: The Introduction of Multimedia in Education*. Aldershot: Ashgate.

Lobet-Maris, C. (1997), 'European policy in multimedia development', first report for the cross-national comparative analysis, mimeo, FUNDP Namur.

Lobet-Maris C., with van Bastelaer, B. and Cammaerts, B. (1999) 'On the role of government in the information society', in B. Cammaerts and J.-C. Burgelman (eds), *Paving the Way for a New Public (tele)Communication Policy*, VUB press.

Lobet-Maris, C. and van Bastelaer, B. (eds) 1999 *Digital Cities: SLIM Integrated Study Final Report*, mimeo – Faculte Universite de Notre Dame de la Paix, Namur: CITA. (http://www.info.fundp.ac.be/~cita/publications/SLIM/). An edited version appears in French translation, as van Bastelaer, B., Henin, L., Lobet-Maris, C., and Eveno, E. (2000), *Villes virtuelles: entre Communauté et Cité: analyse de cas*, Paris: L'Harmattan.

McCracken, G. (1988) *Culture and Consumption: New Approaches to the Symbolic Character of Consumer Goods*, Bloomington, IN: Indiana University Press.

Mackay, H. and Gillespie, G. (1992) 'Extending the social shaping of technology approach: ideology and appropriation', *Social Studies of Science*, 22, pp. 685–716.

Mackay, H., Carne, C., Beynon-Davies, P., and Tudhope, D. (2000) 'Reconfiguring the user: using rapid application development', *Social Studies of Science*, 30 (5), pp. 737–757.

MacKenzie, D. and Wajcman, J. (eds) (1985) *The Social Shaping of Technology: How the Refrigerator Got Its Hum*, Milton Keynes: Open University Press.

McLaughlin, J., Rosen, P., Skinner, D. and Webster, A. (1999) *Valuing Technology: Organisations, Culture and Change*, London: Routledge.

McLoughlin, I. (1999) *Creative Technological Change: the Shaping of Technology and Organisations*, London: Routledge.

Mansell, R. E. (1993) *The New Telecommunications: A Political Economy of Network Evolution*, London: Sage.

Marvin, C. (1988) *When Old Technologies were New: Thinking about Electric Communication in the Late Nineteenth Century*, New York: Oxford University Press.

Mick, D. G. and Fournier, S. (1998) 'Paradoxes of technology: consumer cognizance, emotions and coping strategies', *Marketing Science Institute Working Paper*, Cambridge, MA: MSI.

Miles, I. (1990) *Home Telematics: Information Technology and the Transformation of Everyday Life*, London: Pinter.

Molina, A. (1989) *The Social Basis of the Microelectronics Revolution*, Edinburgh: Edinburgh University Press.

Molina, A. (1992) 'Competitive strategies in the microprocessor industry: the case of an emerging versus an established technology', *International Journal of Technology Management*, 7 (6/7/8) special issue on the Strategic Management of Information and Telecommunication Technology, pp. 589–614

Molina, A. (1994) 'Insights into the successful generation of a large-scale European initiative: from mis-alignment to programmatic alignment in the build up of sociotechnical constituencies', in R. Mansell, (ed), *The Management of Information and Communication Technologies: Emerging Patterns of Control*, London: ASLIB, pp. 90–120.

Monaco, J. (2000) *How to read a film*, Milton Keynes: Open University Press.

Morley, D. and Silverstone, R. (1990) 'Domestic communications: technologies and meanings', *Media, Cutlure and Society*, 12 (1), pp. 31–55.

Morris, C. R. and Ferguson, C. H. (1993) 'How architecture wins technology wars', *Harvard Business Review*, 71 (2), pp. 86–96.

Mourik, R. (2001) 'Multimedia and education as marketing strategy' in M. van Lieshout, T. Egyedi, and W. E. Bijker, (eds), *Social Learning Technologies: The Introduction of Multimedia in Education*, Aldershot: Ashgate, pp 63–82.

Murdock, G., Hartmann, P. and Gray, P. (1992) 'Contextualising home computing: resources and practices', in R. Silverstone, and E. Hirsch, (eds), *Consuming Technologies: Media and Information in Domestic Spaces*, London: Routledge.

Myervohld, N. (1999) Transcript of interview given to BBC Radio 4, *In Business*, 11 November.

Negroponte, N. (1995) *Being Digital*, New York: A. A. Knopf and London: Hodder and Stoughton.

Nelson, R. R. and Winter S. (1982) *An Evolutionary Theory of Economic Change*, Cambridge, MA: Belknap Press.

Nicoll, D. W. (2000) 'Users as currency: technology and marketing trials as naturalistic environments', *The Information Society*, 16 (4), pp. 303–10.

Noble, D. (1979) 'Social choice in machine design: the case of automatically controlled machine tools', in A. Zimbalist, (ed), *Case Studies on the Labour Process*, New York: Monthly Review Press, pp. 18–50.

Nonaka, I. and Takeuchi, H. (1995) *The Knowledge Creating Company*, Oxford: Oxford University Press.

Nordli, H. (2001): 'From "Spice Girls" to Cyber Girls? The role of multimedia in the construction of young girls fascination for and interest in computers', in M. Lieshout, T. M. van Egyedi, and W. E. Bijker (eds), *Social Learning Technologies: The Introduction of Multimedia in Education*, Aldershot: Ashgate.

Norman, D. (1988) *The Psychology of Everyday Things*, New York: Basic Books.

Norman, D. (1990) *The Design of Everyday Things*, London: MIT Press.

Norman, D. (1992) *Turn Signals are the Facial Expressions of Automobiles*, Reading, MA: Addison-Wesley.

Norman, D. A. (1998) *The Invisible Computer: Why Good Products Can Fail, the Personal Computer Is so Complex, and Information Appliances are the Solution*, Cambridge, MA: MIT Press.

Oakley, B. (1990) 'Look back in Alvey: why support for R&D is not enough', Edinburgh PICT Working Paper, no. 18, Edinburgh University.

Oudshoorn, N. and Pinch, T. (2003) *How Users Matter. The Co-Construction of Users and Technologies*, Cambridge, MA: MIT Press.

Oudshoorn, N., Rommes, E. and Stienstra, M. (2004) 'Configuring the user as everybody: gender and design cultures in information and communication technologies', *Science, Technology and Human Values*, 29 (1), pp. 30–63.

Pacey, A. (1983) *The Culture of Technology*, Oxford: Blackwell.
Perez, C. (1983) 'Structural change and the assimilation of new technologies in the economic and social system', *Futures*, 15, pp. 357–75.
Pierre, J. and Peters, B. G. (2000) *Governance, Politics and the State*, Basingstoke: Macmillan.
Pinch, T. and Bijker, W. (1984) 'The social construction of facts and artefacts: or how the sociology of science and the sociology of technology might benefit each other', *Social Studies of Science*, 14 (3), pp. 399–441.
Pollock, N. and Cornford, J. (2001) 'Customising industry standard computer systems for universities: ERP systems and the university as a "unique" organisation' paper presented to Critical Management Studies Conference 2001 Manchester School of Management, UMIST, 11–13 July.
Pollock, N. and Cornford, J. (2003) *Putting the university on-line: information, technology and organisational change*, Milton Keynes: Open University Press.
Pollock, N., Procter, R. and Williams, R. (2003) Fitting standard software packages to non-standard organisations: the 'biography' of an enterprise-wide system, *Technology Analysis* and *Strategic Management,* 15 (3), pp. 317–332,
Porter, M. (1990) *The Competitive Advantage of Nations,* London: Macmillan.
Preston, P. (ed) (1999) *Cultures and Content: SLIM Integrated Study: Final Report*, mimeo, Department of Communication Studies, Dublin City University: Dublin (http://www-rcss.ed.ac.uk/SLIM/private/I-studies/pp/pp.html
Preston, P. (ed) (2000) *New Media and Society*. Themed Section: Content is king? Culture, community and commerce, guest ed. Paschal Preston, 2 (3), pp. 259–334.
Preston, P. (2001) *Reshaping Communications*: *Technology, Information and Social Change*, London and New Delhi: Sage.
Preston, P. and Kerr, A. (2001) 'Digital media, nation-states and local cultures: the case of multimedia "content" production', *Media, Culture and Society*, 23, pp. 109–131.
Procter R. N. and Williams, R. (1996) 'Beyond design: social learning and computer-supported cooperative work: some lessons from innovation studies', in D. Shapiro, M. Tauber and R. Traunmueller (eds), *The Design of Computer-Supported Cooperative Work and Groupware Systems,* Amsterdam: North Holland, pp. 445–64.
Procter, R. N., Williams, R. and Cashin, L. (1999) 'Social learning and new strategies for the management of innovation in multimedia', in R. Slack, J. Stewart and R. Williams (eds), *The Social Shaping of Multimedia,* COST A4, European Commission DGXIII, Luxembourg: Office for Official Publications of the European Community, pp. 155–69.
Quintas, P. (ed) (1993) *Social Dimensions of Systems Engineering: People, Processes, Policies and Software Development*, New York and London: Ellis Horwood.
Qvortrup, L. (1999) 'Interactive multimedia art: the socio-aesthetic shaping of multimedia', in R. Slack, J. Stewart and R. Williams (eds), *The Social Shaping of Multimedia*, Cost A4, European Commission, Luxembourg: Office for Official Publications of the European Community, pp. 219-44.
Rip, A., Misa, T. J. and Schot, J. (eds) (1995) *Managing Technology in Society: The Approach of Constructive Technology Assessment*, London and New York: Pinter.
Rip, A. (1995) 'Introduction of new technology: making use of recent insights from sociology and economics of technology', *Technology Analysis and Strategic Management*, 7 (4), pp. 417–31.
Rip, A., and Schot, J. (1999) 'Anticipation on contextualization: loci for influencing the dynamics of technological development', forthcoming in D. Sauer, and C. Lang

(Hrsg.), *Paradoxien der Innovation. Perspektiven sozialwissenschaftlichter Innovationsforschung*, Frankfurt and New York: Campus Verlag. Proceedings, workshop Verbund sozialwissenschftliche Technikforschung, München, December 1998.

Rogers, E. M. (1983) *Diffusion of innovations*, New York: The Free Press.

Rommes, E. (2002) 'Worlds Apart: exclusion-processes in DDS', in M. Tanabe, P. van den Besselaar and T. Ishida (eds), *Digital Cities II; Computational and Sociological Approaches* (vol. LNCS 2362, pp. 219–32), Berlin and Heidelberg: Springer.

Rommes, E., van Oost, E and Oudshoorn, N. (2001) 'Gender in the design of the digital city of Amsterdam', in E. Green and A. Adams (eds), *Virtual Gender. Technology, Consumption and Identity Matters*, London: Routledge, pp. 241–62.

Rosen, Paul (1993) 'The social construction of mountain bikes: technology and postmodernity in the cycle industry', *Social Studies of Science*, 28 (3), pp. 479–513.

Rosenberg, N. (1979) *Perspectives on Technology*, Cambridge: Cambridge University Press.

Rosenberg, N. (1982) *Inside the Black Box: Technology and Economics*, Cambridge: Cambridge University Press.

Rosenberg, N. (1994) *Exploring the Black Box: Technology, Economics and History*, Cambridge: Cambridge University Press.

Rosenbloom, R. S. and Cusumano, M. A. (1987) 'Technological pioneering and competitive advantage: the birth of the VCR industry', *California Management Review*, 29, pp. 55-76.

Rossel, P. (1999) 'Multimedia in Switzerland', in R. Williams and R. S. Slack (eds), for *Europe Appropriates Multimedia*, Trondheim: Norwegian University of Science and Technology.

Russell. S., and Williams, R. (2002a) 'Social shaping of technology: frameworks, findings and implications of policy, with glossary of social shaping concepts', in K. H. Sørensen and R. Williams (eds), *Shaping Technology, Guiding Policy: Concepts, Spaces and Tools*, Edward Elgar: Aldershot, pp. 37–132.

Russell. S., and R. Williams (2002b) 'Concepts, spaces and tools for action? Exploring the policy potential of the social shaping of technology: perspective', in K. H. Sørensen and R. Williams (eds), *Shaping Technology, Guiding Policy: Concepts, Spaces and Tools*, Edward Elgar: Aldershot, pp. 133–54.

Schmutzer, R. (1999) 'The "social construction" of new media characteristics and effects', in Slack, Stewart and Williams (eds), *The Social Shaping of Multimedia*, COST A4, European Commission DGXIII, Office for Official Publications of the European Community: Luxembourg, pp. 209–18.

Schneider, V. (1997) 'Different roads to the information society? Comparing U.S. and European approaches from a public policy perspective', in H. Kubicek, W. Dutton and R. Williams (eds), *The Social Shaping of Information Superhighways: European and American Roads to the Information Superhighway*, Frankfurt: Campus Verlag, pp. 339–58.

Schneider, V. (2000) 'Evolution in cyberspace: the adaptation of national videotext systems to the Internet', *The Information Society*, 16 (4), pp 319–28.

Schneider, V, Charon, J.-M., Miles, I., Thomas, G. and Vedel, T. (1991) 'The dynamics of videotex development in Britain, France and Germany: a cross-national comparison', *European Journal of Communication*, 6 (2), pp. 187–212.

Schon, D. A. (1983) *The Reflective Practitioner: How Professionals Think in Action*, London: Temple Smith.

Schot, J. and Rip, A. (1996) 'The past and future of constructive technology assessment', *Technology, Forecasting and Social Change*, 43 (1), pp. 251–68.

Schumm, W. and Kocyba, H. (1997) 'Recontextualisation and opportunities for participation: the social shaping of implementation', in C. Clausen and R. Williams (eds), *The Social Shaping of Computer-Aided Production Management and Computer-Integrated Manufacture*, Luxembourg: European Commission, pp. 49–62.

Silverstone, R. (1991) 'Beneath the bottom line: households and information and communication technologies in an age of the consumer' *PICT Policy Research Papers*, No. 17, Swindon: Economic and Social Research Council.

Silverstone, R. (1994) *Television and everyday life*, London: Routledge.

Silverstone, R. and Hirsch, E. (eds) (1992) *Consuming Technologies: Media and Information in Domestic Spaces*, London: Routledge.

Silverstone, R., Hirsch, E. and Morley, D. (1992) 'Information and communication technologies and the moral economy of the household', in R. Silverstone and E. Hirsch, (eds), *Consuming Technologies. Media and Information in Domestic Spaces*, London: Routledge.

Slack, R. (2000a) 'Learning in Cable School', in M. van Lieshout, T. Egyedi, and W. E. Bijker (eds), *Social Learning in Multimedia: Education*, Aldershot: Ashgate.

Slack, R. S. (2000b) 'Community and technology: social learning in CCIS', in M. Gurstein, (ed), *Community Informatics: Enabling Communities with Information and Communication Technologies*, Hershey, PA: Idea Group.

Slack, R., Stewart, J. and Williams, R. (eds) (1999) *The Social Shaping of Multimedia*, COST A4, European Commission DGXIII, Luxembourg: Office for Official Publications of the European Community.

Slack, R. S. and Williams, R. A. (2000) 'The dialectics of place and space: On community in the "Information Age"', *New Media and Society*, 2 (3), pp. 313–34.

Smith, S. and Wield, D. (1987) 'New technology and bank work: banking on I.T. as an "organizational technology"', in L. Harris, (ed), *New Perspectives on the Financial System*, London: Croom Helm.

Soete, L. et al. (1997) *Building the European Information Society for Us All: Final Policy Report of the High-Level Expert Group*, Directorate-General for employment, industrial relations and social affairs, Brussels: European Commission, europa.eu.int/comm/employment_social/ knowledge_society/buildingen.pdf.

Sørensen, K. H. (1994) 'Technology in use. Two essays on the domestication of artefacts', *STS Working paper*, 2/94, Trondheim: Centre for Technology and society.

Sørensen, K. H. (1994a) *The Car and Its Environments: The Past, Present and Future of the Motor car in Europe*, proceedings of COST A4 Workshop, Trondheim, Norway, 6–8 May 1993, COST Social sciences, vol. 2. Luxembourg: European Commission DGXIII Science Research and Development.

Sørensen, K. H. (1994b) 'Adieu Adorno: the moral emancipation of consumers', in A.-J. Berg and M. Aune (eds), *Domestic Technology and Everyday Life – Mutual Shaping Processes*, COST Social sciences vol. 1, Luxemburg: European Commission DGXIII Science Research and Development, pp. 157–69.

Sørensen, K. H. (1996) 'Learning technology, constructing culture. Socio-technical change as social learning', *STS working paper*, no 18/96, Trondheim: University of Trondheim: Centre for Technology and Society.

Sørensen, K. H. and Berg, A.-J. (eds) (1992) *Technologies and Everyday Life: Trajectories and Transformations*, proceedings from a workshop in Trondheim,

28–29 May 1990, Report No. 5, Oslo: Norwegian Research Council for Science and the Humanities.

Sørensen, K. H. and Spilker, H. (1999) *Multimedia in the Home: An analysis of SLIM findings*, Annex 4 to Social Learning in Multimedia – final report, http://rcss.ed.ac.uk/research/slim.html

Sørensen, K. H. and Williams, R. (eds) (2002) *Shaping Technology, Guiding Policy: Concepts, Spaces and Tools*, Aldershot: Edward Elgar.

Sørensen, K. H., Aune, M. and Hatling, M. (1996) 'Against linearity. On the cultural appropriation of science and technology', *STS Working Paper*, 8/96, Trondheim: Norwegian University of Science and Technology.

Sørensen, K. H. and Stewart, J. K. (eds) (2002) Digital divides and inclusion measures: A review of literature and statistical trends on gender and ICT, senter for Teknologi og Sumfunn, Trondheim, NTNU, Norwegian University of Science and Technology Working Paper 59–2002.

Spacek, T. R. (1997) 'How much interoperability makes an NII?', in H. Kubicek, W. Dutton and R. Williams (eds), *The Social Shaping of Information Superhighways: European and American Roads to the Information Superhighway*, Frankfurt: Campus Verlag, pp. 69–77.

Spilker, H. and Sørensen, K. H. (2000) 'A ROM of one's own or a home for sharing? Designing the inclusion of women in multimedia', *New Media and Society*, 2 (3), pp. 268 - 87

Spilker, H. and Sørensen, K. H. (2002) 'Feminism for profit? Public and private gender politics in multimedia', in K. H. Sørensen and R. Williams (eds), *Shaping Technology, Guiding Policy: Concepts, Spaces and Tools*, Aldershot: Edward Elgar.

Star, S. L. and Griesemer, J. R. (1989) 'Institutional ecology, "translations", and boundary objects: amateurs and professionals in Berkeley's museum of vertebrate zoology 1907–1939', *Social Studies of Science*, 13, pp. 205–28.

Steenhoudt, F. (1997) 'De nieuwe cyberpalen: gebruiksonvriendelijk (Techno Antwerpen Deel 2)', in *De Morgen* – Metro, 3 March, p. 8.

Stewart, J. (1999) 'Cafematics: communities in the computer', in M. Gurstein (ed), *Community Informatics*, Toronto: Idea Group.

Stewart, J. (1999a) The Web meets the TV: users and the innovation of interactive television. Television of the Future – or: The Future of Television? in C. Toscan and J. Jensen (eds) Aalborg: Aalborg University Press, pp. 25–66.

Stewart, J. (2001) 'Encounters with the Information Society: Personal and social issues in the appropriation of new media products in everyday life: adoption, non-adoption, and the role of the informal economy and local experts', unpublished PhD thesis, Research Centre for Social Sciences, University of Edinburgh.

Suchman, L. A. (1987) *Plans and Situated Actions: The Problem of Human-Machine Communication*, Cambridge: Cambridge University Press.

Tait, J. and Williams, R. (1999) 'Policy approaches to research and development: foresight, framework and competitiveness', *Science and Public Policy*, 26 (2), pp. 101–12.

Thomas, G. and Miles, I. (1990) *Telematics in Transition* London: Longman.

Vedel, T. (1994) 'Introduction à une socio-politique des usages', in A. Vitalis (sous la direction de), *Médias et nouvelles technologies: Pour une socio-politique des usages*, Rennes: Editions Apogée, pp. 13–34.

Vedel, T. (1996) 'Information superhighway policy in France: the end of high tech colbertism', in B. Kahin, and E. Wilson (eds), *National Information Infrastructure Initiatives, Vision and Policy Design*, Cambridge, MA: MIT Press.

Vincenti, W. G. (1994) 'The retractable airplane landing gear and the northrop "anomaly": variation-selection and the shaping of technology', *Technology and Culture*, 35, pp. 1–33.

Wajcman, J. (1991) *Feminism confronts technology*, Cambridge: Polity.

Webster, J. (1990) *Office Automation: The Labour Process and Women's Work in Britain*, Hemel Hempstead: Harvester Wheatsheaf.

Webster J. and Williams, R. (1993) 'Mismatch and tension: standard packages and non-standard users', in P. Quintas (ed), *Social Dimensions of Systems Engineering: People, Processes, Policies and Software Development*, Hemel Hempstead: Ellis Horwood, pp. 179–96.

West, J. (1996) 'Utopianism and national competitiveness in technology rhetoric: the case of Japan's information infrastructure', *The Information Society*, 12, pp. 251–72.

Whipp, R. (1985) *Innovation and The Auto Industry: Product, Process and Work Organization*, London: Pinter.

Williams, R. (1990) *Television, Technology and Cultural Form*, London: Routledge.

Williams, R. (1993) *Information Technology in Organisations*, report to European Commission DGIII Industrial Policy Directorate, IT Strategy Unit, as part of a study on the *Socio-Economic Aspects of a European Information Space*. Available as *Edinburgh PICT Working Paper* No 54 (1994) Edinburgh University/RCSS, Edinburgh.

Williams, R. (ed) (1995) *The Social Shaping of Inter-Organisational Network Systems and Electronic Data Interchange*, proceedings of international conference, COST A4, European Commission DGXIII, Luxembourg, pp. vi, 216

Williams, R. (1997) 'The social shaping of information and communications technologies', in H. Kubicek, W. Dutton and R. Williams (eds), *The Social Shaping of Information Superhighways: European and American Roads to the Information Society*, New York and Frankfurt: Westview/St. Martins Press.

Williams, R. (1997a) 'Globalisation and contingency: tensions and contradictions in the mutual shaping of technology and work organisation', in I. McLoughlin, and D. Mason (eds), *Innovation Organizational Change and Technology*, London: International Thompson Business Press, pp. 170–85.

Williams, R. (1999) 'The national appropriation of multimedia', in R. Williams and R. Slack (eds), *Europe Appropriates Multimedia: A Study of the National Uptake of Multimedia in Eight European Countries and Japan*, Senter for Teknologi og Samfunn (Centre for Technology and Society), Report No. 42 Norwegian University of Science and Technology, Trondheim: Norway. ISSN 0802 3581 42, pp. 1–19.

Williams, R., Faulkner, W. and Fleck, J. (eds) (1998) *Exploring Expertise*, Basingstoke: Macmillan.

Williams, R. and Edge, D. (1996) 'The social shaping of technology', *Research Policy*, 25, pp. 865–99.

Williams, R. and Slack, R. (eds) (1999) *Europe Appropriates Multimedia: A study of the National Uptake of Multimedia in Eight European Countries and Japan*, Senter for Teknologi og Samfunn (Centre for Technology and Society), Report No. 42 Norwegian University of Science and Technology, Trondheim: Norway, ISSN 0802 3581 42.

Williams, R. (2000) 'Public choices and social learning: the new multimedia technologies in Europe', *The Information Society*, introduction to special issue on ICT development and use in Europe, 16 (4), pp. 251–62.

Winner, L. (1980) 'Do artifacts have politics?' *Daedalus*, 109 (1). Reprinted in *The Social Shaping of Technology*, eds D. McKenzie and J. Wajcman (1985), London: Open University Press.
Winston, B. (1998) *Media Technology and Society, a History: From the Telegraph to the Internet*, London and New York: Routledge
Woolgar, S. (1991) 'Configuring the user: the case of usability trials', in J. Law (ed), *A Sociology of Monsters. Essays on Power, Technology and Domination*, London: Routledge, pp. 57–102.
Wynne, B. (1995) 'Technology assessment and reflexive social learning: observations from the risk field', in A. Rip, T. J. Misa and J. Schot (eds), *Managing Technology in Society: The Approach of Constructive Technology Assessment*, London and New York: Pinter, pp. 19–36.

Index